# UNDERSTANDING INTUITION

# UNDERSTANDING INTUITION

## A Journey in and Out of Science

**LOIS ISENMAN**
Women's Studies Research Center
Brandeis University
Waltham, MA, United States

ELSEVIER

ACADEMIC PRESS
An imprint of Elsevier

Academic Press is an imprint of Elsevier
125 London Wall, London EC2Y 5AS, United Kingdom
525 B Street, Suite 1800, San Diego, CA 92101-4495, United States
50 Hampshire Street, 5th Floor, Cambridge, MA 02139, United States
The Boulevard, Langford Lane, Kidlington, Oxford OX5 1GB, United Kingdom



**Library of Congress Cataloging-in-Publication Data**
A catalog record for this book is available from the Library of Congress

**British Library Cataloguing-in-Publication Data**
A catalogue record for this book is available from the British Library

ISBN: 978-0-12-814108-3

For information on all Academic Press publications visit our website at https://www.elsevier.com/books-and-journals



*Publisher:* Nikki Levy
*Acquisition Editor:* Emily Ekle
*Editorial Project Manager:* Carly Demetre
*Production Project Manager:* Priya Kumaraguruparan
*Designer:* Miles Hitchen

Typeset by TNQ Books and Journals

*To all those who desire to maintain the wonder of the ineffable even as they struggle to pierce its veil.*

# CONTENTS

# ACKNOWLEDGMENTS

To begin, I would like to acknowledge my scientific mentors, who had such an important role in this journey. Deep appreciation to the late Paulo Dice for helping me see that my cognitive style is heavily weighted towards intuition, and his enthusiastic exploration with me of what intuition means, especially for scientists. Heartfelt thanks also to Stephen Rothman for earlier introducing me to bidirectional parallel interactive interconnection and for the privilege of working in his laboratory as a graduate student. Deep appreciation to Florence Ladd and the 1994–95 Bunting Institute fellows for their support of my fledgling attempts to delve deeper into intuition, and to Shula Reinharz and the Women's Studies Research Center of Brandeis University for providing me a stimulating and warm environment during most of the time I was writing this book. Special thanks to Mei-Mei Ellerman, Marcie Tyre, Nancer Ballard, Mary Mason, Rosie Rosenzweig, Nurit Eini-Pindyck, and Margaret Gullette of the Women's Studies Research Center Memoir Group for reading and critiquing many of these chapters. Appreciation to Paul Cooper for his Zen wisdom, to John Haule for his insight about Jung, and to Marta Sinclair for her skillful championing of intuition in the academic community and her support of my efforts. Thanks to the impressive Brandeis undergraduates who kept reinspiring me with our work on the "The Bridge: A Science and Spirituality Resource." Thanks also to Emily Elke and April Farr for taking a chance on this unusual book, and to Emily, Carly Demetre, Priya Kumaraguruparan, Anita Vethakkan, and many others at Elsevier, for their skillful help. I cannot express enough appreciation to my partner Joel for his sustaining love and friendship and his support for my work; and the same applies to my brother to my brother Paul for being there, and for countless thought-provoking conversations about intuition over the years. Finally, deep appreciation to my mother, from whom I acquired my intuitive side, and to my father, from whom I acquired my scientific bent.

# INTRODUCTION

*Understanding Intuition: A Journey in and Out of Science* explores the underlying the biological/cognitive mechanisms that likely account for intuition and also the first-person experience of intuition, resonating the two whenever possible. Intuition has recently emerged from its cloud of suspicion and become the focus of intense scientific study. Although the gap between scientific and first-person approaches to understanding remains substantial, rapid progress is occurring. It is an increasingly alive gap, pregnant with potential.

Intuition is largely grounded in unconscious knowledge and understanding. During the past 2 decades, experimental evidence has revealed that the unconscious mind is much more active in normal functioning than many had thought possible. Recent work suggests that neural activity predicting voluntary movement can be detected as early as 10 s before subjects become aware of the urge to move.[1] A number of studies show that the prefrontal cortex, which previously was thought to be active only in association with consciousness, can influence behavior in response clues that do not appear in consciousness.[2,3] Current research also extends the perceived boundaries of unconscious cognition into the realm of abstract symbolic representation, which is an important prerequisite for sophisticated intuition. Recent studies, for example, have shown that the unconscious mind can understand simple phrases,[4,5] distinguish between compound adjectives such as "not good" and "very good,"[6] and also perform simple arithmatic.[4,7]

A book bridging the science and experience of intuition is timely for several reasons in addition to the accelerating pull between them. Many of the probable cognitive and biological mechanisms underlying intuition are now well-studied areas of scientific investigation. Among these are implicit learning, parallel-distributed information processing, complex systems theory, the role of emotion in cognition, mirror neurons, and the likely centrality of metaphor in the development and extension of human intelligence. We can already comprehend much about the cognitive potential of the unconscious mind from a scientific perspective.

Moreover, a book that brings together the science and experience of intuition is needed at this juncture because intuition is increasingly studied in the laboratory. Understanding it is viewed more and more as a scientific endeavor. Many cognitive scientists (perhaps paradoxically seeming to lay readers) appear to have damped their curiosity about the experience of intuition, especially its more sophisticated aspects—or put it on hold.

As a scientist, I appreciate the empiric urge behind the desire to frame understanding intuition as a completely scientific task. I also share in the excitement about current scientific inroads into the cognitive workings of the unconscious mind. Yet studying intuition is unlike studying most other phenomena that are the focus of scientific investigation. To explore it only at the scientific level is like trying to understand a language only by learning its grammar and vocabulary, not by speaking or reading it. To really comprehend intuition, it is necessary to enter into the experience—to approach it from the inside—in addition to exploring its underlying science.

*Understanding Intuition: A Journey In and Out of Science* bears witness to the surprising power of the unconscious mind, while grounding this knowledge in scientific understanding. The scientific and experiential approaches belong together, even though their match is far from perfect at present. The 2 perspectives bootstrap on each other, with each opening up additional room for the other. The more deeply we explore how they fit together, the better we will be able to understand the potential power of the human mind.

Science is a slow incremental process. The task of integrating empirical and experiential understanding of intuition will likely go on for many years, and it may never be complete. Nonetheless I am hopeful that the future will strengthen, not overturn, most of what is written here. At least so far, this has been true.

This book has taken many years to publish: the project began in the late 1990s. In the meanwhile, many of the views that I argue are foundational to understanding intuition have become more visible in the cognitive science literature. Some that were on the fringe years ago when the first draft was written (such as the importance and potential subtlety of metacognitive feelings) have moved toward the mainstream. Others that were nowhere in sight then (such as the view that working memory can sometimes function below awareness) are beginning to affect the field. Also, some of the unanswered questions that earlier drafts highlighted as needing investigation (such as whether implicit knowledge can build on explicit knowledge) are now being investigated. I have had to rework some of the material several times to reflect important new studies, but the core ideas have remained relatively unchanged.

My intuition, I must quickly add, deserves the credit to the extent that some of these ideas were initially before their time. Likewise, my confidence in them, in addition to their holding up thus far, rests on my intuition having taken the lead in my investigation, and in an unusual way. I began to explore intuition seriously as a science scholar at the Bunting Institute of Radcliffe College in 1994–95. In the beginning of the year, my plan was to try to follow my intuition in action as I went about my scientific work. Much to my surprise, instead I started having some powerful intuitive experiences about the process of intuition itself. These striking encounters with my unconscious mind also helped direct my reading in the scientific literature. Along with the empirical and theoretical material, they have an essential place in the book.

## A BLINK OF THE EYE

One of these foundational experiences has proved an especially useful guide to understanding intuition; it echoes throughout the book. The following is from a vignette I first published in 1997 in an article entitled "Towards an Understanding of Intuition and Its Role in Scientific Endeavor."[8] Its topmost level has already demonstrated its mettle and become part of the common lexicon:

> *Early in my investigation of intuition, a colleague surprised me by asking for an impromptu definition of intuition. After a moment of searching, the phrase a blink of the eye unexpectedly came to mind. This didn't seem to answer the question, so I looked again. This time a subliminal physical sense of the experience of intuition as a rapid closing and opening somewhere in my chest accompanied the mental experience of seeing a generic signal go down and then up on a graph.*

*Understanding Intuition: A Journey In and Out of Science* uses this metaphor for unconscious cognition and intuition (subsequently popularized by Malcolm Gladwell[9]) along with the intuitive encounter in which it came to me to help pierce the veil of consciousness. In conjunction with the scientific and other experiential material, it reveals layer after layer of insight about what intuition is and how it works.

When I had no ready answer to my friend's daunting question, "What is intuition?" instinctively I turned my attention inside. In response, my mind below my awareness apparently called up a number of its different aspects (such as its swiftness, its physical resonance, and subliminal quality) and unified them all with the deceptively simple, old-fashioned idiom, "a blink of the eye." Intuition occurs when a number of different strands of information coalesce below awareness into a novel and significant piece of understanding. In *prototypic* experiences of intuition, we end up knowing something important in an instant without knowing how, and know it to be true.

But this was not a prototypic experience of intuition. I did not know what the phrase "a blink of the eye" had to do with intuition, or whether it had any validity. Nevertheless, rather than a handicap, this was a gift, because looking inside again brought to awareness two additional images. The three together validated the first and provided an indelible glimpse of some of the generally hidden workings of the unconscious mind.

## INTUITION AS A PROCESS

We do not always become aware of intuitions in an instant. They often depend on a complicated interaction between the unconscious and conscious mind. Sometimes we might sense a vague edge or corner of an idea that feels compelling and work on it both above and below awareness before it yields a clear insight.

At other times, a number of seemingly unconnected (and sometimes apparently contradictory) images or strands of information present themselves to the conscious mind around a topic. On the surface, for example, little seems to link the verbal phrase "a blink of the eye" to the subtle kinesthetic image or feeling of an increase and decrease of tension in my chest, or to the visual image of a signal going down and up on a graph or oscilloscope. Yet somehow, I realized that the pattern of motion was the same in all three. This made me recognize "a blink of the eye" as a valid characterization of intuition even though I understood neither the individual images nor their common pattern of motion. Shortly afterwards the conversation moved on to another topic. Only some months later, while preparing a talk, did I begin to see the relevance of the idiom "a blink of the eye" as well as the other images to intuition and to appreciate some of the meaning of the common "down–up" pattern of motion.

Intuition often begins with *not knowing*. Recognizing not knowing something creates the space for its eventual knowing. It also humbles consciousness so that it is more willing than usual to hand control over to the unconscious mind and to wait and see what emerges. The blink-of-the-eye encounter turns back on itself and helps capture at both the process and content levels the experiential core of many important intuitions that emerge only over time.

These *slow-cooking* intuitions can integrate a number of more or less relevant-seeming strands of information. They can also encode multiple different levels of understanding that eventually converge at a deeper level of integration to form a rich and resonant understanding. Instead of yielding immediate illumination, intuitions that take us deep down into the hidden aspects of the mind often first generate a feeling of conscious bewilderment. This is one layer of meaning of the *down* of the down–up motif unifying the three different images that appeared from my unconscious in my first few attempts to answer the question, "What is intuition?" Slow-cooking intuitions can also require conscious effort to harvest their bounty. Moreover, a number of successive cycles of bewilderment and illumination may occur, leading to an ever more integrated and richer understanding. My hope is that the book will provide the ingredients for a slow-cooking understanding of intuition as a whole.

## THE INTERCONNECTED ELEMENTS

Chapter 1 provides an overview of intuition by exploring three challenging but basic questions about this elusive mental activity: (1) What is the scientific status of unconscious cognition, including high-level unconscious cognition? (2) What distinguishes intuition from ordinary cognition, much of which also just pops into mind? (3) What is the relationship between scientific and artistic intuition?

The next four chapters focus on current scientific understanding about unconscious intelligence and unconscious perception. This material grounds intuition in the evolution of biological intelligence. I also include a small amount of experiential subject matter to build more direct links between scientific and experiential ways of knowing.

Chapter 2 reviews seminal work in the field of implicit learning. It shows that humans learn many patterns of regularity in the environment without effort and without awareness of learning anything. This body of knowledge makes the provocative suggestion that the unconscious mind is better than the conscious one at picking up many complex patterns of interrelationship. Implicit learning, along with other kinds of tacit knowledge, or knowledge not readily available to awareness, can influence behavior from below awareness or appear woven into an intuition.

Chapter 3 examines intuitive cognition, the interwoven or parallel interactive way of processing information that characterizes much unconscious cognition and is at the core of most intuition. It contrasts this *logic of the whole* to the serial logic that characterizes conscious thought at its most rigorous. The chapter also considers brain laterality, or how the two sides of the brain tend to process information differently, and how this relates to parallel interactive processing.

Chapter 4 explores the brain and perception. It begins with connectionism, or neural network information-processing architecture, which is modeled on the parallel interactive way in which neurons interact with each other. Several examples illustrate how it can mimic many basic human perceptual skills. After a primer on complexity theory, I offer a more complicated model based on the work of Walter Freeman, which is grounded in complexity theory and allows for the much richer connectivity of the human brain. The model gives a central place to emotional factors and focuses on the internal meaning that objects and events have for the brain. It proposes that perception occurs largely below awareness: a process that I argue is governed by an intuitive, or implicit, self. The model can begin to account for the stability of ordinary perception as well its potential dramatic reorganization during intuition.

The final chapter of the section focuses on emotion, an evolutionary adaption that helps organisms pay attention to what is important to their survival and well-being. Emotion provides the driving force for all behavior and cognition, and thus for intuition. I highlight landmark studies by Damasio that suggest most emotions are bodily responses based on past experience and occur largely below awareness. From there they help guide perception and intuition through bodily feelings or their mental derivatives. These studies have elucidated the important somatic component to intuition, and they have also been essential to the academic acceptance of intuition as a real phenomenon. The chapter also considers a fast emotional pathway that may allow emotional and motivational influences a say in how deeply the brain processes various stimuli.

The middle section of the book explores the mental mechanisms that bring hidden understanding to consciousness. It includes chapters on mental images, embodied experience and imagery, and feelings of truth. These chapters move freely between scientific and experiential material.

Chapter 6 focuses on the connection between mental imagery, intuition, and imagination. I argue that mental imagery is required to bring novel and sophisticated unconscious knowledge and understanding to consciousness. Imagination generates the specific images, sometimes fanciful and sometimes not, that convey these potential breakthrough intuitions to awareness.

Chapter 7 examines the special place that embodied experience and embodied mental imagery has in intuition. It is structured around an unusual intuitive encounter I had many years ago that helps bring home the foundational role of bodily experience in our understanding of the external world. The chapter also considers mirror neurons, which show that this shadowy correspondence between inner and outer experience based on bodily experience is not unique to humans. I argue that for those who favor intuitive perception, even very abstract nonconscious knowledge can become translated via imagination back into bodily sensations or body-based imagery.

Chapter 8 considers *feelings of rightness*, or cognitive evaluative feelings. It centers on another intuitive encounter that illustrates a number of different kinds of feelings of rightness and how they function in insight formation and intuition. This unusual experience also points to several different layers of cognitive functioning below awareness. This chapter, like Chapter 4, argues for the existence of an intuitive self that organizes our experience below awareness. The last part of the chapter focuses on the subtle feeling of truth that tends to accompany intuition at its most profound.

Chapter 9, the final chapter, explores the intersection of intuition and spirituality by asking, "Who are we?" I argue that our ego, which is formed in interaction with the world, is grounded in survival. It is often the source of intuition, because its needs guide perception below awareness. Yet many influences, from Internet technology to climate change, increasingly bind humanity into a complex system in which we all influence each other. Our Being, which we experience consciously in moments of personal Presence, transcends the boundaries of ego and recognizes itself as part of a larger whole. Its demand for more inclusive and far-seeing meaning making is the wellspring of more profound intuition, and with it greater authenticity, power, and wholeness.

## TURNING INTUITION ON ITSELF

I have tried to make the book a good read. This in part means making sure that all of the information is clearly presented and accessible. I hope I have succeeded. Even so, some may find it hard to hold all of the different threads together at a conscious level, because they include scientific and experiential perspectives, as well as a number of strands that weave back and forth. The unconscious mind is much better suited than the conscious mind for the complex parallel interactive processing required.

The inability to integrate or understand at a conscious level is not necessarily a disadvantage. Instead, it can be a prelude to a fuller encounter with intuition if it results in consciously turning toward the unconscious mind. Intentionally turning towards the hidden aspects of mind eventually will provide a direct experience of the "down–up" movement that characterizes many of the deepest intuitions. It will also provide strong evidence for the integrating power of the unconscious mind.

Intuition at almost every level of understanding depends on wholeness through complex interconnection. In part by interweaving different elements as well as different perspectives, intuition allows us to move beyond what is known into the unknown. The more disjunctive the entwining strands, the deeper the ensuing journey. Likewise, by exploring intuition from both an objective and subjective perspective, by exploring different empirical approaches to understanding, as well as different kinds of intuitive experiences, *Understanding Intuition: A Journey In and Out of Science* tries to focus intuition on itself. Along with conveying the potential power and wisdom of the unconscious mind, turning intuition on itself also begins to evoke the ineffable: the nature of biological intelligence, and with it something about who we humans are.

## REFERENCES

1. Soon CS, Brass M, Heinze H-J, Haynes J. Unconscious determinants of free decisions in the human brain. *Nat Neurosci* 2008;**11**:543–5.
2. van Gaal S, Ridderinkhof R, Scholte SS, Lamme VAF. Unconscious activation of the prefrontal no-go network. *J Neurosci* 2010;**30**:4143–50.
3. Lau HC, Passingham RE. Unconscious activation of the cognitive control system in the human prefrontal cortex. *J Neurosci* 2007;**27**:5805–11.
4. Sklara AY, Levy N, Goldstein A, et al. Reading and doing arithmetic unconsciously. *Proc Natl Acad Sci U S A* 2012;**109**:19614–9.
5. van Gaal S, Naccache L, Meuwese JDI, et al. Can the meaning of multiple words be integrated unconsciously. *Phil Trans R Soc B* 2014;**369**(1641):20130212. https://doi.org/10.1098/rstb.2013.0212.
6. Armstrong AM, Dienes Z. Subliminal Understanding of Negation: unconscious control by subliminal processing of word pairs. *Conscious Cogn* 2013;**22**:1022–40.
7. Ric F, Muller D. Unconscious addition: when we unconsciously initiate and follow arithmetic rules. *J Exp Psychol Gen* 2012:141222–6.
8. Isenman L. Towards an understanding of intuition and its importance in scientific endeavor. *Perspect Biol Med* 1997;**40**:395–403.
9. Gladwell M. *Blink: the power of thinking without thinking*. New York: Little Brown; 2005.

# CHAPTER 1

# Some Basic Questions

## Contents



Understanding intuition requires entering the poorly illuminated realm of the unconscious mind. To prepare for the journey ahead, this initial chapter sheds some light on three basic but thorny issues about unconscious intelligence and intuition: (1) What is the current scientific status of unconscious intelligence? (2) What distinguishes ordinary everyday thought, which tends just to pop into mind, from the often startling and much more rare experience we call intuition? (3) What is the relationship between scientific and artistic intuition?

Before proceeding, a preliminary comment about what the word *intuition* means: Thinking about intuition is so difficult in part because it refers to several different but interrelated aspects of certain mental activity. At one level, *the content level*, intuition refers to insights that pop into mind without effort and reorient understanding in important, novel, and sometimes radical ways. At another level, *the processing level*, it refers to a specific way of integrating information that generally occurs below awareness and supports direct knowing, or knowing without conscious thinking. At yet another level, *the evaluative level*, it refers to a largely unconscious mode of evaluation


*Understanding Intuition*
ISBN 978-0-12-814108-3
https://doi.org/10.1016/B978-0-12-814108-3.00001-4

that allows us to judge whether our own or others' ideas are true. Each aspect of intuition implies that the unconscious mind possesses surprising intelligence.

## THE SCIENTIFIC STATUS OF UNCONSCIOUS INTELLIGENCE

There are two very different scientific views about whether the unconscious mind is capable of the kind of sophisticated intelligence required by many accounts of intuition. One, a theoretical perspective, is strongly supportive. The other, an experimental perspective, thus far is much more equivocal.

Evolutionary theory is entirely consistent with the idea that the unconscious is capable of considerable intelligence. Intelligence developed for eons in lower animals before consciousness appeared. Unconscious intelligence can account for most of the complex behavior and cognition that occur in other higher animals. The evolutionary perspective suggests that unconscious intelligence is not what requires explaining because it is the norm. Rather the oddity in evolution, what needs explaining, is consciousness itself.

Some cognitive scientists see consciousness not only as a late addition to animal intelligence but also as one of marginal necessity. Called epiphenomenologists,[1] they understand consciousness to be like a local radio announcer who reads the news but has little or no role in deciding what stories to report or what details to include. Conscious processes are determined completely by underlying unconscious ones, and consciousness may have no causal role in thinking or behavior. For epiphenomenologists, unconscious higher intelligence is a given.

### Who's in Charge?

Important early support for the epiphenomenologists' notion of consciousness came from the provocative results of a series of experiments by neurophysiologist Benjamin Libet[2–4] in the 1970s and 1980s. As he recorded electrical activity from the scalp of his subjects, Libet had them move their wrist whenever they felt the urge. He also asked them to mark the instant they first felt the urge by noting the position of a bright red dot that swept around a large clock face once every 2.6 s. He found that *the readiness potential*, a characteristic electroencephalogram, or EEG, activity associated with pending voluntary movement, occurred about a third of a second before the subjects felt the conscious urge to move. If the brain can signal forthcoming movement before subjects experience the conscious impulse to act, the movement cannot be caused by a conscious intent. Libet found that such an action is voluntary, or consciously willed, in the limited sense that the conscious mind can veto it in the roughly 150 ms between experiencing the urge to move and beginning the movement.[3]

A more recent experiment by Soon and colleagues[5] extended Libet's results using functional magnetic resonance imaging (fMRI) and statistical pattern recognition techniques. When subjects felt the urge to move, they pressed one computer key with a finger of the right hand or another with a finger of the left hand. The researchers were able to extract predictive patterns of brain activity from specific areas of subjects' prefrontal and parietal cortex signaling the hand they would use as early as 10 s before they were conscious of deciding. Also the researchers were able to extract patterns from the supplementary motor area predicting the time of movement about 5 s before the decision became conscious.

The readiness potential apparently occurs late in the buildup of brain activity leading to this kind of voluntary movement. The decision to move appears to depend on prior activity in a number of regions of the brain at different times. However the patterns observed were only 60% predictive, which suggests that other factors also could have a role in the decision to move. In another experiment, Soon and colleagues found that whether a subject chose to add or subtract two numbers was predicted by brain activity 4 s before they were aware of deciding.[6] Again consciousness cannot cause the decision if brain activity prior to consciousness can predict the outcome.

Another of Libet's findings is equally unsettling.[4] In patients undergoing brain surgery, he found that stimulating a specific region of the somatosensory cortex (where sensations register in the brain) and maintaining it for half a second generates a sensation of touch on the wrist. In some experiments, 200 ms after stimulating the patients' cortex, he also touched them briefly on the other wrist. Common sense says that they should feel the sensation generated by direct cortical stimulation first because it occurred first, because the signal from the other wrist had to travel to the cortex, and because cortical processing of the second signal should also have taken about half a second. Yet they felt the external stimulation first, not the direct cortical stimulation.

The key to these perplexing results likely lies in the different pattern of electrical activity the two kinds of stimulation evoke in the cortex.[4,7] The actual touch on the wrist generates a brief peak of electrical activity about 20 ms after the stimulation as well as an additional extended peak beginning roughly 80 ms later. With direct cortical stimulation, in contrast, only the one extended peak of electrical activity occurs. After additional control experiments,[4] Libet concluded that the brain takes about half a second (indicated by the extended peak of activity) to integrate any input and to determine whether it is important enough to come to consciousness. If it is, the brain "backdates" it to the time of the brief initial peak that marks its arrival at the cortex. Subjects *feel* the touch on the wrist first because direct stimulation of the cortex does not generate this time stamp, and therefore it is not backdated.

These experiments suggest that what is *now* for the conscious mind is actually a half a second ago. At some level, we know that something like this has to be true because we tend to respond involuntarily to sensory information that might be critical to survival before it comes to awareness. We find ourselves removing our hand from the stove and then realizing the stove is hot, or we startle and then recognize what is startling us. Taken together, Libet's findings suggest that much of the time we respond (or our mind prepares to respond) to sensory information before it is conscious—or fully conscious.[8]

On occasion, when we are on the edge of our abilities and precise timing can make a difference in the success or grace of our actions, the unconscious mind seems to take over completely. On a difficult ski run, it might make a series of choices and execute the required movements, each without the conscious mind being aware of a need to choose. We experience the paradoxical feeling of being completely in control even as we may realize we are functioning outside conscious control. We are *in the zone.*

Libet's studies shed substantial doubt on our sense that our conscious mind is in charge.[9,10] In his book *The Illusion of Conscious Will*, Daniel Wegner collected a number of additional lines of evidence for this view.[11] For example, studies of posthypnotic suggestion as well as studies of split-brain patients (in whom the right and left hemisphere were surgically divided) have helped to reveal the human tendency to make things up after the fact to rationalize experience or behavior. Wegner's own work showed that subjects could be tricked into judging acts that were actually controlled by someone else as their own, using a computerized version of a Ouija board and a graduate student "confederate." The effect was evident when the confederate forced a stop at a certain place *as much as 30 s before* the experimenter actually cued the subject to stop there. Wegner proposed that both the experience of will, or intention, and the action that is generally assumed to be its result are caused by prior unconscious processes.[11,12]

Most people who study consciousness accept that there is a broad general continuity in the development of intelligence within evolution. They also subscribe to the view that most lower-level cognitive processing happens below awareness. For the most part, we do not have to think about the meaning of what others say, look for appropriate memories, or think about how to construct grammatically correct sentences. Processes outside awareness take care of all of this. Evolution favors unconscious information processing because it is fast and can deal with many different tasks at the same time. In contrast, conscious processing is much slower and has a much more limited capacity. As a result, it is prone to bottlenecks. Imagine how inefficient (to say nothing of dangerous) life would be if we continually had to waste conscious resources on the activities just mentioned.

Although most cognitive scientists subscribe to this view, many do not go as far as those who view consciousness as only an epiphenomenon of underlying unconscious processes. In contrast to epiphenomenologists, they believe that unconscious information processing is not of sufficient complexity to account for all human thought, or they are reluctant to pass judgment on the issue. Instead they believe, or are at least open to the idea, that the complex intelligence that is so central to being human resides to a greater or lesser extent in the conscious mind and conscious abilities.

Whether or not this is the case, the experience of consciousness is so primary for us humans, it is where we must begin. This is true even for those who hold the theoretical position that consciousness does not really do anything except read a script prepared by underlying unconscious processes. All of our inquiry about the world and ourselves necessarily employs the conscious mind. Our most basic assumption has to be that at least our own consciousness is real, even if we believe that consciousness is the oddity in evolution that requires explaining.

In contrast, by definition, complex hidden intelligence is unavailable to experience; by the same token, it is much more difficult to demonstrate in humans than conscious intelligence. Therefore it has a more questionable empirical status. For a whole group of scientists, the formal assumption with which they begin their work must be that sophisticated unconscious intelligence does not exist, despite a theoretical climate in cognitive science that is favorable to its existence. Experimentalists who work on the question must collect evidence that shows that the capacity for complex unconscious understanding does (or does not) exist in humans.

## A Brief History of the Study of Unconscious Cognition

Despite the difficulty inherent in collecting evidence for unconscious cognition in humans, the case for it has been building slowly for a long time. In the 17th century, Leibniz recognized that conscious experience is influenced by unconscious perceptions. In the 19th century, Helmholtz realized that the unconscious mind must use inference in addition to sensory clues to construct the perceptual world that is given to experience.[13] Late in the 19th century, the unconscious came under scientific scrutiny as the repository of repressed emotional feelings. Freud and others used hypnosis to demonstrate the existence of an unconscious emotional life in patients with psychiatric disorders.[14] They showed that the patients' symptoms often improved when repressed conflicts were brought to consciousness.

In the 20th century, cognitive scientists discovered a host of neurological conditions that provided evidence for a cognitive unconscious. One of the earliest characterized was *blindsight*.[15] Because of damage to part of the primary visual cortex, patients with blindsight can experience complete blindness on one side of the visual field. Yet they can still process and act on certain visual information presented there. They might be able to grasp an object correctly, or determine whether a bar is horizontal or vertical, even though they think they are only guessing.

The late neurologist Oliver Sacks, in his 1985 book *The Man Who Mistook His Wife for a Hat*, documented for a lay audience a number of bizarre neurological problems in patients that reflect the malfunction of different unconscious cognitive processing mechanisms.[16] The man for whom the book is named has *Capgras Syndrome*. Capgras patients cannot recognize family members, although visual recognition of famous people may be normal, and voice recognition of family members may also be normal.[16,17] Disorders such as Capgras that create havoc in a very specific aspect of perception highlight the large number of unconscious information-processing steps that shape conscious experience below awareness.

The evidence for complex or higher-level cognition by the unconscious is weaker than for lower-level cognition. Although it is fragmentary and uneven, it provides tantalizing glimpses of the sophisticated cognitive potential of unconscious mental activity, as many of these chapters will show. One well-established area of study, reviewed in Chapter 2, that points to an important role for the unconscious mind in higher-level cognition is implicit learning.[18,19]

Young children learn to speak their native language, which requires combining words to create sentences they have not heard before, by registering without effort or awareness the grammatical regularities or patterns that characterize the speech of others. Implicit learning explores the ability of the mind to direct behavior on the basis of complex regularities it extracts below awareness from experience. Experiments with artificial grammars as well as other implicit learning studies suggest that the unconscious mind is better suited than the conscious mind to picking up many, if not most, complicated patterns.

Yet some studies point to limitations in the ability of humans to collect data below awareness. Certain experiments show that the presence of distractors or having subjects perform difficult additional tasks at the same time tends to reduce implicit learning.[20,21] A minimum level of selective attention or perhaps processing capacity may be required.[22]

### *Subliminal Stimulation*

The ability of subliminal stimuli—stimuli that do not reach consciousness—to *prime,* or enhance, a subsequent response was another early line of evidence in support of unconscious cognition. A series of experiments by Zajonc,[23] beginning in the 1960s, showed that exposing subjects subliminally to an unfamiliar picture made them more likely to prefer it when it was subsequently shown together at the visible level with another unfamiliar picture. Subliminal pictures are readily processed by the brain, and their influence can last up to several minutes. Yet this is a low-level sensory effect.

A more recent goal of many subliminal perception studies has been to determine whether the brain can decode the meaning of subliminal words. That is, can subliminal stimuli result in the higher-level memory traces, or *representations*, required to process words at the semantic, or meaning, level? In an influential study, Greenwald and colleagues[24] exposed subjects extremely briefly to emotion-laden written words (such as peace or joy, or sadness or gloom) under conditions that ensured they did not appear in consciousness. They found a facilitated response to subsequent visible words that were emotionally congruent and a retarded response to those that were emotionally incongruent. Yet they saw no evidence for unconscious semantic understanding of two-word phrases.

Another landmark study, by Dehaene and colleagues,[25] argued that both Arabic numbers and number words (such as *one* and *four*) that are prevented from reaching consciousness nonetheless can undergo semantic decoding as well as further processing. They asked subjects to respond with one hand when a number word or Arabic number presented was below five, and with the other hand when it was above. Without the subjects' knowledge, they also flashed a subliminal number or number word before each visible stimulus, some that should generate a congruent response and some an incongruent one. They found that the subliminal primes not only influenced subjects' response time in the expected direction, but also initiated subtle neural and motor responses on the appropriate side. Moreover, mixing Arabic and number words between the subliminal primes and the visible targets did not affect the results.

However critics pointed out that because the subliminal words and numbers were the same as the visible ones in both experiments just described, the results could be explained instead by stimulus–response learning from the visible trials.[26] This, as well as the possible one-word restriction, left open the possibility of serious limitations in the unconscious abilities of humans as they relate to words and symbols, which have such an important role in human knowledge. Since then a number of critical repeats of the last experiment, along with a spate of other studies, argue strongly for semantic decoding below awareness[26] (although some investigators may still remain unconvinced).[27] One study, for instance, showed some crossover priming between subliminal written and spoken number words, which requires the formation of abstract high-level unconscious representations.[28]

Using clever techniques, researchers are beginning to find evidence that the unconscious can indeed process two or more words as well as engage in other high-level mental activity previously assumed to occur only in consciousness. One report showed that the unconscious can distinguish between incoherent and coherent three-word written phrases presented subliminally: for instance, between "I ironed coffee" and "I made coffee."[29] Another showed that it can distinguish between compound adjectives presented subliminally, such as "not good" and "very good."[30] Several reports also showed that the unconscious can perform at least simple arithmetic.[29,31]

Many studies use the enhanced brain imaging and recording techniques that first appeared in the 1990s—such as functional magnetic resonance imagining, or fMRI. Exploration of unconscious intelligence increased dramatically with the development of these powerful tools that allowed researchers to follow changes in brain activity as well as potentially correlate them with behavior. One especially illuminating study from the late 1990s used fMRI to document that pictures of angry faces that were flashed too quickly to register in awareness nonetheless activated the amygdala, the part of the brain that registers danger, triggering an emotional response.[32] This implies that some intuitions about others can come from emotional facial signals that are too brief for the conscious mind to register. More powerful imaging techniques and experimental paradigms are constantly emerging, providing progressively better access to hidden mental activity. They are allowing researchers to extract and decode increasingly subtle unconscious mental activity.

These new tools have also extended the recognized boundaries of unconscious intelligence into the executive functioning of the mind, which most before thought was tightly linked to consciousness. One critical early study showed that when either of two cues that were previously associated with different tasks at a conscious level (identifying if a word was bisyllabic or if it referred to a concrete object) are presented subliminally before the same visible word, they initiate brain activity in different higher-level regions of the brain.[33] A series of studies also show that when the unconscious mind is given a subliminal cue asking it to *inhibit* a subsequent response, it can slow a consciously-triggered subsequent action, and the same frontal lobe brain network is activated whether the cue is subliminal or visible.[34] Moreover the slowing occurs in a flexible manner that cannot be explained by stimulus–response conditioning.[35] However the inhibitory activity is much weaker with subliminal stimuli. This is also true in most studies of subliminal perception because subliminal stimuli generate weaker memory traces.

The empirical case for *sophisticated* unconscious intelligence remains rudimentary and uneven despite the significant progress summarized here. What appeared true for many years has changed. Yet these changes, as groundbreaking as they are, represent only first steps. On the basis of reasoning well-grounded in scientific principles, some cognitive scientists seriously entertain the idea that consciousness is a mere epiphenomenon of unconscious processes. Yet others who begin from equally sound but different scientific principles, despite their hard work, find only preliminary empirical evidence supportive of the kind of sophisticated unconscious intelligence required by higher-level intuition.

Science is a dynamic process. The gulf between these two extreme points of view will likely continue to shrink as increasingly sensitive and sophisticated brain imaging and recording techniques as well as novel experimental paradigms extend the evidence for higher-level nonconscious cognition. One very recent paper, for example, suggests that the unconscious mind can perceive a story from a series of subliminal pictures.[36]

As this brief history of subliminal semantic priming emphasizes, broad conclusions drawn from specific experiments are likely to be revised with the development of new experimental approaches, as well as more nuanced manipulation of old ones. Arguably, sweeping *negative* conclusions based on one experiment, or even a series of them, are especially subject to revision. Meanwhile, the scientific case for sophisticated unconscious cognition and for intuition is stronger than those who focus exclusively on the results from studies using subliminal stimuli might conclude.

### *Intuition Without Subliminal Stimuli*

Studies using subliminal stimulation have been critical to legitimizing the notion of unconscious intelligence and expanding its perceived boundaries into the abstract symbolic realm as well as into executive functionig. Yet whatever the limits of subliminal perception turn out to be, they do not necessarily delineate the reach of unconscious intelligence or of intuition. There are a number of additional sources of knowledge upon which unconscious intelligence can draw.

Some potentially conscious information may remain tacit or implicit, because conscious attention was elsewhere when it was acquired.[37] Compared with subliminal stimuli, potentially conscious but unattended stimuli form stronger memory traces and undergo deeper information processing. Equally important, the unconscious mind can extract *higher-level knowledge* from experience that is fully attended and readily available to awareness, as Chapter 2 on implicit learning documents. Likewise, a number of experiments argue that when there are many pieces of conscious information to integrate, the unconscious mind can sometimes make better decisions than the conscious mind: for example, deciding which car is most practical based on a large number of fixed criteria.[38]

An extensive body of work in social sciences also argues that experiences that are conscious nonetheless can influence many aspects of cognitive function and behavior below awareness. It reveals, for example, that simply by decoding anagrams or unscrambling sentences using words related to achievement, rudeness, or old age, subjects can be primed to act more achievement oriented, rude, or elderly.[39,40] Together these studies suggest that the unconscious mind has a pervasive influence on all mental functioning. Commenting in 2000 on the already vast literature on the "automaticity" of thought and behavior, Bargh and Fergerson[41] concluded that "social interaction, evaluation and judgment, and the operations of internal goal[s] ... can all proceed without the intervention of conscious acts of will and guidance of the process." The social science evidence has grown greatly since then.[42,43] The unconscious mind is influenced by a wide range of information and experience, some which maybe wholly or partially available to awareness.

## THE DIFFERENCE BETWEEN ORDINARY THOUGHT AND INTUITION

To differentiate intuition from ordinary thought, it is useful to take a close look at the indistinct boundary separating them. Initially this might seem to blur the line between the two even more. However, eventually it should provide a framework for distinguishing them as well as understanding the role of unconscious processes in each.

First consider what is meant by *ordinary thought*. The word *thought* is often used loosely to refer to the contents of consciousness, especially its verbal contents. Yet at other times, it is reserved for the deliberate mental activity involved in solving a difficult problem. Ordinary thought that occurs in the absence of conscious effort is *automatic* or *passive*, and the more deliberate and effortful variety is *active*, or *directed*.

Automatic and active thought are often considered the product of two different mental systems, typically called *System 1* and *System 2*.[44] The two-systems view is useful for differentiating habitual thought or, more broadly, going with the first thing that comes to mind (System 1), from reasoned analysis (System 2). Intuition is often seen as synonymous with the products of System 1. All intuition does share some important characteristics with the kind of automatic mental activity that governs much of human thought and behavior. However higher-level intuition tends to oppose automatic or habitual thought and instead reorients it in novel and important ways. Also unlike automatic or habitual thought, higher-level intuition often begins with not knowing and may take time to develop. To accommodate this view of intuition as a complex and innovative cognitive capacity, I find it more useful to place automatic and active thought on a continuum that is anchored on one side by daydreaming and on the other by rational analysis. Intuition draws from both sides of the continuum.

### Automatic Thought

Most of what occupies consciousness does not consist of thought in the more deliberate sense of the word. The contents of consciousness tend to consist of a stream of images derived from the external world, bodily sensations, fragments of past experience, and fantasy in various combinations—as the writing of James Joyce illustrates. Sometimes this automatic mental activity is composed of chains of mental images stimulated by one or more of above sources. For the most part, we are only dimly aware, if at all, of what triggers our mental associations.

Imagine the following scenario. Driving home after work, Leo notices a billboard picturing a cruise ship, followed by the thought of a tennis match he recently watched on television. This, in turn, triggers a thought about an old college friend and tennis partner, which is followed by a poignant, bittersweet feeling at having lost touch with so many old friends.

After this chain of associations, Leo might wonder why he feels this unaccustomed way. Playing the chain backward will not necessarily reveal what caused his mind to gravitate toward his lost friend and the poignant, bittersweet feeling. Starting from the billboard, his associations might have gone in so many other directions as well.

### Emotion/Motivation and Associations

Passive thought depends to a large extent on emotional factors that remain below awareness. Freud and Jung demonstrated that free associations are guided in part by repressed emotional conflicts. However according to more contemporary emotion researchers, most emotional responses, not just repressed conflicts, tend to be hidden from awareness.[45,46] Like more conscious emotional reactions, unconscious emotional responses strongly influence the flow of unfocused mental experience. In the scenario just presented, an unperceived feeling in response to something much more current and closer to home than losing touch with an old friend may have steered Leo's chain of associations toward the poignant, bittersweet feeling: for instance, his son and frequent tennis partner leaving home for college.

Emotion is a powerful evolutionary adaptation that generates coordinated involuntary bodily adjustments to help biological organisms attend to what is important for their survival and well-being. It guides behavior and mental activity by mediating between an animal's needs on the one hand and what the environment has to offer on the other. Emotion automatically tethers what is valuable to an organism (which is the most basic level of meaning) to biology. In large part toward this end, in humans it channels moment-to-moment mental attention even when the mind is not focused on an external task. The logically disjunctive and frequently fanciful mental associations that tend to result, what we call daydreaming, often have an important place in intuition, as will become clear in a while.

The *associative logic* that guides daydreaming is characteristic of most unconscious intelligence. It differs from the sequential logic of conscious intelligence at its most rigorous. Instead it mimics the parallel interactive way the nervous system has of making connections. Unconscious intelligence depends in large part on the capacity of the unconscious mind under the guidance of the emotional and motivation systems to integrate multiple strands of information as well as multiple levels of understanding simultaneously to create meaningful knowledge of the world. (Of course, with fanciful and sometimes creative associations, the meaning is not always available to the conscious mind.) This nonsequential, interactive way of synthesizing information is variously called association, massively parallel processing, connectionism, parallel distributed processing, parallel interactive processing—or what I like to call it, *intuitive processing*.

### Empirical Association

Another kind of associative connection making, in addition to fanciful or creative association, is characteristic of both ordinary mental function and many aspects of intuition. *Empirical* associations register multiple spatial and temporal aspects of the world. They allow the mind to recognize below awareness, and without effort the regularities, or co-occurrences, that characterize the environment. Sometimes called subsymbolic information processing,[47] empirical associations account for our ability to recognize without thinking the characteristic patterns of features that comprise objects and events. They allow us to "just know" that the small furry creature sleeping by the fireplace in a new friend's house is a large cat, not a small dog, even if it happens to be partially obscured by a chair.

Empirical association making is also responsible for much of expert knowing, such as the direct knowing, or *just knowing*, that underlies much of the work of medical specialists. Like Mark Twain, who was once apprenticed to a steamship captain, experts come to "read the river."[48] This depends on multiple strands of interrelated information coalescing below awareness into a familiar pattern, steered by current needs and priorities that do not themselves necessarily come to consciousness. The correct pattern tends just to pop into awareness, or the appropriate behavior just occurs.

Like ordinary direct or just knowing, intuitions often pop into the mind suddenly. They also depend on the convergence of different interrelated strands of information below awarness into a meaningful pattern. Moreover they are directed by emotional priorities, and thus goals, that may remain below awareness. However unlike ordinary direct knowing, which brings into awareness already familiar patterns, intuitions bring into awareness new patterns that can alter thought or behavior in significant and sometime radical ways. For instance, a medical specialist might suddenly realize (perhaps when thinking of something else) that a number of puzzling cases over the past year with different presenting symptoms all reflect the same novel underlying pattern.

### *Intuition Often Combines Empirical and Creative Associations*

Intuition readily combines the creative and the empirical-based aspects of associative connection making. Fanciful or creative associations and their close cousins, metaphors and analogies, are often pivotal to intuition (see the next section). Spurred by implicit, or unconscious knowledge along with goals that may not be conscious, they can bring to awareness important novel patterns that result in breakthrough understanding.

An incipient intuition grounded in unconscious knowledge (complete or incomplete) can drive the flow of mental associations from below awareness. The mental imagery available in the fringe of awareness sometimes provides a partial window into the process. Often a specific image or chain of images will seem to trigger the appearance of an intuition in awareness.

Empirical studies lend support to the potential productivity of the apparently unfocused mind. Early investigation into daydreaming suggested that it is driven largely by goals that are blocked from fruition.[49] A more recent brain imaging study confirmed that the executive regions of the brain can be surprisingly active during mind wandering, which is consistent with the idea that the daydreamer can be engaged in achieving underlying goals.[50] Mind wandering often seems to dissipate the emotion associated with frustrated needs and goals. Yet in association with implicit or tacit knowledge, it can help further blocked goals (or in some cases, more realistic alternative ones).

In light of the cognitive potential of daydreaming, imagine a somewhat longer associative chain for Leo. In this scenario, after the picture of a cruise ship on a billboard leads to thoughts of an old friend with whom he has lost contact, he remembers a striking story his old friend told him. As he thinks about the twist with which it ends, he senses a pattern on the fringe of awareness. Soon afterward, the solution to a difficult problem he worked on several years before without success pops into his mind. He recognizes that it is also important to the success of his current project.

In the unlikely event that Leo were to focus on the chain of associations that preceded the breakthrough, he might sense a similarity between the solution and the pattern that immediately preceded it. He might even sense a connection between this pattern and the prominently displayed cruise ship logo on the billboard. Chapter 4 explores a model that can begin to account for the capacity of the mind to integrate all of these different levels of meaning-making simultaneously.

## Active Thought

Deliberate or active mental activity can also participate in intuition. It serves conscious goals and recruits focused attention. These goals might be concrete and require a series of specific actions, for example, when fixing a car. However, they may instead be more open-ended, such as figuring out the potential theoretical implications of surprising experimental results, or how to juggle unexpected houseguests on an already overly busy weekend.

The same kind of information processing that is responsible for automatic thought can also explain much of the influence of the unconscious mind on deliberate thought. Below awareness, it integrates all relevant factors together and hands the next thing to say or the next link in the argument to the conscious mind. As the saying goes, "How do I know what I think until I see what I say?"[51]

Heuristics, the mental shortcuts that take the place of rational analysis in much of ordinary thought,[52] also tend to occupy a middle ground between passive and active mental activity. Some consider heuristics to be synonymous with intuition because they contrast with rational though. Yet, unlike intuition, they do not generate novel content. Rather they substitute something cognitively simple for something more complex.[53] The *availability heuristic*, for example, judges the frequency of an event by the ease or fluency with which it comes to mind. Like passive thought, heuristics tend to depend on the convergence of multiple interactive cues below awareness. But like deliberate thought, they often respond to conscious goals and focused attention.

Deliberate conscious analysis, in contrast to heuristics, employs rigorous reasoning. However this does not mean that it excludes unconscious factors.

A mathematical proof, for instance, may initially seem amenable to a variety of different approaches. Yet success may ultimately depend on using a certain one. When we make the right choice in situations like this, we tend to attribute it to good luck. But it is at least equally likely that the unconscious mind had a part in this apparent good fortune.[54]

What might seem to be a completely rigorous conscious assessment of a situation may also depend heavily on unconscious influences. The mathematician and physicist Henri Poincaré[55] stressed that reason is only a tool of intuition. Likewise, biologist Jonas Salk,[56] whose work helped eradicate polio, argued that reason is what we use to convince ourselves of what our intuition has already told us. Studies of judgment and decision making lend some support to this view.[45,57]

Implicit or tacit knowledge can work to influence not only the formation of scientific hypotheses but also experimental design and even the analysis of data. Influences that function below consciousness may direct a scientist to place more emphasis on certain aspects of a situation than others.[58] Below awareness, she might sense a pattern in preliminary data and then design a series of experiments that happen to exploit this trend. She might even come up with an entirely different conscious justification for the experimental plan.

At other times, unconscious insight about a pattern might partially come to consciousness. Those aspects of the data that reflect this pattern might seem particularly salient to the investigator, who begins his conscious reasoning from there. He does not necessarily know he is viewing the data through an unconscious filter; rather he may feel he is simply seeing *what is there*. Indeed, who is to say he is not?

Whether or not they come to awareness, intuitive assumptions complicate the issue of scientific objectivity. The historian of science Gerald Holton[59] examined the famous *oil-drop* experiment by Stanley Millikan, which showed that electronic charge increases and decreases in discrete steps. Using the original laboratory notebook, Holton documented that Millikan, who was already convinced that the electron was a discrete unit of charge, discarded all but 58 of his 140 trials. (In the same chapter, Holton also revealed that another researcher who performed a similar experiment but used all of the data he collected concluded erroneously that the charge on the electron varies continuously, a conclusion also in line with the researcher's own prior bias.) Millikan had technical reasons for discarding many of his points. However at least some of his technical judgments were likely influenced below awareness by his bias about how the data should look. Even in science, directed thought is the tool of its premises and goal—sometimes for the better and sometimes for the worse.

## Premises and Goals Tend to Anchor All Thinking in Unconscious Influences

Premises as well as goals link the self-consistency of reason with useful knowledge. Arguably all premises rest *ultimately* on intuition. Einstein wrote, "The supreme task of the physicist is to arrive at those universal laws from which the cosmos can be built up by pure deduction." Yet, he continued, "There is no logical path to these [elementary] laws, only intuition resting on sympathetic understanding of experience."[60]

With both everyday thought and intuition, intuitive processes have an important role in turning sensory and cognitive experience into the assumptions and then into the premises from which the conscious mind begins to reason or plan a course of action. Some of these premises are of course consensual, some we learn from other people and then confirm for ourselves, and others we take on good authority. All the rest we put together on our own, often on the fly. We depend on the unconscious for the most part to understand the meaning of what we have just heard, seen, or read. The hidden aspects of mind, with their fine discrimination abilities, capacity to integrate large amounts of information, and speed, are especially well suited to ground the reasoning or action that may follow.

Both ordinary everyday active thought and the active thought serving intuition are anchored in unconscious influences. The difference is that active thought serving intuition incorporates significant novel assumptions dependent largely on unconscious knowing into its premises. These premises then propel thought along novel lines that can account for all of the relevant information in a new and more coherent manner than the older view. Einstein, for example, fashioned the theory of special relativity on the premise that the speed of light is invariant, an idea based largely on his sense of the harmony of the universe.[58] The success of the theory in accounting for the physical world in a more consistent way than classical physics in turn justified to him the use of this intuitive premise.

The novel assumptions behind breakthrough intuitions do not always come to awareness, which explains why someone whose directed thought is grounded in unconscious knowledge can mistake it for reason. If an intuition seems to follow naturally from its premises, it will appear logically compelling *even when these premises depend on hidden assumptions.* The researcher who based on a relatively obscure similarity feels certain that two systems will respond in the same way, may below awareness recognize a deeper underlying pattern connecting them. Only when trying to communicate his thinking to someone else might he realize that more than logic is at play here.

### Intuition at Its Richest Seems to Come From a Unified Aspect of Mind

The experience of intuition at its most fertile often combines elements from both ends of the mental continuum. On occasion the goal-directed attention associated with deliberate thought moves from ordinary mental activity toward what appears to be a more unified layer of mind. Mental focus becomes unusually penetrating yet effortless as the passive and active aspects of mind work synchronously. Creative associations, direct knowing, and reasoned analysis function together toward a common and often unanticipated end.

Yet such a turn away from ordinary awareness is only minimally voluntary. This profound mental sensing is instead something that just happens to us. Although this heightened state tends to last for only a short time, at its peak it brings to awareness a surge of interrelated images that together illuminate a novel mental vista or greatly enhance an already familiar one. For an instant we are seeing, knowing, and reasoning all at once, and the distinction among them all but disappears. On these rare occasions the veil that clouds ordinary vision falls away and we comprehend with a depth and completeness—with a wholeness—that is generally unavailable to us.

## THE RELATIONSHIP BETWEEN SCIENTIFIC AND ARTISTIC INTUITION

The historian of science Gerald Holton[58] divided scientific endeavor into two parts. His two-tiered view of scientific work helps illuminate the link between scientific and artistic intuition. *Science 1*, or *public science*, is science as it appears in journal articles and eventually in textbooks. It moves from hypothesis to experiment to data and back again in a neat upward spiral. Yet Holton stressed that much of the time, investigators recreate after the fact the thinking that went into their experiments in order to fit this orderly public view of science.

*Science 2*, or *private science*, is science the way it actually happens. Holton used a number of case studies to document the less rational and largely unconscious influences that strongly influence scientific work. These include a scientist's feeling for the subject matter and the cultural zeitgeist of the day. Private science, like artistic endeavor, depends to a large extent on personal and cultural influences that tend to remain mostly below awareness where they may mix with implicit or tacit knowledge.

Like artists, scientists are guided in their work by hard to articulate urges that flow from the convergence of myriad influences and strands of information below awareness. The intuitive mind harvests these impulses to give shape and then voice to what was before unknown and at the same time to confer a feeling of rightness and knowing upon this new knowledge. If necessary, the process may recur at ever greater levels of refinement.

A number of pivotal scientists openly acknowledged their debt to intuition. Mendeleev, who developed the periodic table, and Kekulé, who discovered the structure of the benzene ring, join Einstein in crediting intuition—in fact, unusual intuitive experiences—with the impetus for their critical insights.[61–63] Mendeleev's and Kekulé's breakthroughs were sparked by dreams, and the kernel of the idea for the special theory of relativity came to the adolescent Einstein in the fanciful image of riding on a light beam.

Many other scientists have also written or spoken about the role of intuition in their work. Yet the influence of intuition is not limited to scientists who are willing to acknowledge its impact on their endeavors. Nor is it limited to scientists who are aware of its role in their work. The rational veneer of the conscious mind can hide the influence of the unconscious mind.

Nonetheless the disconnection between the license generally associated with artistic endeavor and the rigor associated with scientific work has not entirely disappeared. One of my first glimpses of intuition in action stressed its poetic nature. Early in my investigation, a friend asked me for a definition of intuition. Completely at a loss, I turned my attention inward and the now famous metaphor for intuition, "a blink of the eye," came to mind (see Introduction).[64,65] In the context of a poem it would not be unusual to think of intuition as "a blink of the eye." However, clearly it will not do as a scientific description of intuition—even though some scientific studies do point to a connection between eye blinks and intuition.[66]

The phrase is, of course, a metaphorical rather than rigorous characterization of intuition. Yet this does not mean that it is irrelevant to understanding scientific intuition. The form, or underlying structure, of this intuitive depiction of intuition highlights the nature of unconscious thought and therefore the link between scientific and artistic intuition. "A blink of the eye" is a powerful metaphor for intuition because it interweaves a number of different characteristics as well as different levels of understanding of intuition. The poetic richness of the image depends on a number of obvious as well as more obscure associations, along with the complex interconnections among them. Its rightness and richness tend to be experienced below awareness even if the conscious mind cannot grasp any of its more apparent meanings. Although my conscious mind in the moment judged the phrase to be a silly answer to the question "What is intuition?" I found it compelling enough to repeat to my friend.

On occasion, fanciful images more suited to poetry than science, such as Einstein's image of riding on a light beam, or Kekulé's dream image of a snake chasing its tail, can serve as advance notice of an incipient scientific insight. The unconscious mind has a proclivity for fanciful metaphors because they encode many different layers of experience and meaning all together. They allow the unconscious to capture and express its richly simultaneous understanding. The densely interwoven texture that typifies forceful metaphors tends to characterize all intuition, be it scientific, poetic, or spiritual.

Science, like poetry, requires complex simultaneous understanding. At a public level it deals in impersonal physical *meanings* rather than the personal or cultural *meanings* that are central to artistic expression. However even public science requires richly layered insight to unify the different strands and levels of observation that go into understanding and confirming understanding of most complex physical phenomena. As we strive to comprehend the physical world at a deeper level, we seek basic theories that will unify more and more physical phenomena. The potential of the unconscious mind for complex simultaneous understanding is well suited to this task.

Although scientific intuition shares certain characteristics with poetic intuition, it is, of course, much less free. Both can join observations derived from different kinds of observation or experience. Yet scientific intuition must preserve potentially relevant distinctions, whereas poetic intuition can merge such distinctions with impunity. In a similar vein, both can combine different levels of understanding. Yet scientific intuition must strive to make them logically cohesive, whereas poetic intuition can ignore the strictures of logic.

This is not to say that scientific intuition is devoid of leaps of creative imagination. It is not limited to amalgamating evidence but it can create more general theories that reconfigure what we already know as well as predict whole classes of phenomena that have not yet come to light. Such is the character of profound scientific thought. Like poetic imagination, scientific imagination can soar above the physical world as we know it. Unlike poetic imagination, however, it is always constrained to take the physical (or mathematical) world into account.

At its richest, intuition appears to come from a part of the mind that is more foundational than the split between the associative thought characteristic of poetry and the rigorous thinking central to science. The metaphors that sometime inspire and guide private science, such as Einstein's image of riding on a light beam, often encode associative as well as rigorous layers of understanding. At a level of mind that consciousness can rarely glimpse, the separation between the subjective and objective realms is less clear than we generally assume.

The task of understanding intuition dovetails to a large extent with what I see as one of the biggest challenges for cognitive scientists in the coming years. It is to understand more fully the biological and cognitive mechanisms that account for the hidden unity along with the overt disunity of what we think of as the objective and the subjective components of human knowledge. Intuition allows us to glimpse the wholeness of human intelligence that lies beneath the dance of the two facets of our knowing.

## REFERENCES

1. For a general discussion of epiphenomenalism see Flanagan O. *Consciousness reconsidered*. Cambridge (MA): MIT Press; 1992.
2. Libet B, Gleason CA, Wright EW, Pearl DK. Time of conscious intention to act in relationship to the onset of cerebral activity (readiness potential): the unconscious initiation of a freely voluntary act. *Brain* 1983;**106**:623–42.
3. Libet B. Unconscious cerebral initiative and the role of conscious will in voluntary action. *Behav Brain Sci* 1985;**8**:529–66.
4. Libet B, Wright EW, Feinstein B, Pearl DK. Subjective Referral of the timing for a conscious sensory experience. *Brain* 1979;**102**:193–224.
5. Soon CS, Brass M, Heinze H-J, Haynes J-D. Unconscious determinants of free decisions in the human brain. *Nat Neurosci* 2008;**11**:543–5.
6. Soon CS, He AH, Bode S, Haynes J-D. Predicting free choice for abstract intentions. *Proc Natl Acad Sci U. S. A* 2013;**110**:5733–4.
7. Edelman G, Tononi G. *A universe of consciousness*. New York: Basic Books; 2000. p. 68. The early peak in response to the external stimuli is localized to one specific region of the cortex. It occurs even in the absence of consciousness, and it varies only with stimulus intensity. The later peak is widely distributed over the cortex, does not occur during anesthesia or slow wave sleep, and it is much more variable, changing with alterations in attention.

8. For a more recent study supporting backdating see Banks WP, Isham EA. We infer rather than perceive the moment we decide to act. *Psychol Sci* 2010;**20**:17–21.
9. For a different interpretation of the readiness potential see Lavazza A. Free will and neuroscience: from explaining freedom away to new ways of operationalizing and measuring it. *Front Hum Neurosci* 2016;**10**:262. https://doi.org/10.3389/fnhum.216.00262.
10. For a critique of Libet's interpretation of his backdating data see Dennett D. *Consciousness explained*. New York: Little Brown; 1991. p. 153–66.
11. Wegner D. *The illusion of conscious will*. Cambridge (MA): Bradford Books; 2002.
12. For an interesting study that lends support to this view see Desmurget M, Reilly KT, Richard N, et al. Movement intention after parietal cortex stimulation in humans. *Science* 2009;**324**:811–3.
13. Kihlstrom JF. *The rediscovery of the unconscious*. 1999. Available from: http://ist-socrates.berkeley.edu/~kihlstrm/rediscovery.htm.
14. For a comprehensive history of the early investigation of the emotional unconscious see Ellenberger H. *The discovery of the unconscious*. New York: Basic Books; 1970.
15. Weiskrantz L. *Blindsight: a case study and its implication*. Oxford: Oxford University Press; 1986.
16. Sacks O. *The man who Mistook his Wife for a Hat, and other clinical tales*. New York: Summit; 1985.
17. Reid I, Young AW, Hellawell DJ. Voice recognition impairment in a blind Capgras patient. *Behav Neurosci* 1993;**6**:225–8.
18. Reber AS. *Implicit learning and tacit knowledge*. New York: Oxford University Press; 1993.
19. French A, Cleeremans A, editors. *Implicit learning and consciousness*. New York: Psychology Press; 2002.
20. Jimenez L, editor. *Attention and implicit learning*Philadelphia: John Benjamins; 2003.
21. Tanaka D, Kiyokawa S, Yamada A, Dienes Z. Role of selective attention in artificial grammar learning. *Psychon Bull Rev* 2008;**15**:1154–9.
22. Lavie N. Distracted and confused? Selective attention under load. *Trends Cogn Sci* 2005;**9**:75–82.
23. Kunst-Wilson WR, Zajonc RB. Affective discrimination of stimuli that cannot be recognized. *Science* 1980;**207**:557–8.
24. Greenwald AG, Draine SC, Abrams RL. Three cognitive markers of unconscious semantic activation. *Science* 1996;**273**:1699–702.
25. Dehaene S, Naccache L, Le Clec'H G, et al. Imaging Unconscious semantic priming. *Nature* 1998;**395**:597–600. [The effect occurred even when Arabic numerals and number words were mixed between prime and target].
26. Kouider S, Dehaene S. Levels of processing during non-conscious perception: a critical review of visual masking. *Philos Trans R Soc Lond B* 2007;**362**:857–75.
27. Newell BR, Shanks DR. Unconscious influences on decision making: a critical review. *Behav Brain Sci* 2013;**37**:1–19.
28. Kouider S, Dehaene S. Subliminal number priming within and across the visual and auditory modalities. *Exp Psychol* 2009;**56**:418–33.
29. Sklara AY, Levy N, Goldstein A, Mandel R, et al. Reading and doing arithmetic nonconsciously. *Proc Natl Acad Sci U. S. A* 2012;**109**:19614–9.
30. Van Gaal S, Naccache L, Meuwese JDI, et al. Can the meaning of multiple words be integrated unconsciously? *Philos Trans R Soc Lond B* 2014;**369**(1641):20130212. https://doi.org/10.1098/rstb.2013.0212.
31. Ric F, Muller D. Unconscious addition: when we unconsciously initiate and follow arithmetic rules. *J Exp Psychol Gen* 2012;**141**:222–6.
32. Whalen JP, Rauch SL, Etcoff NL, et al. Masked presentations of emotional facial expressions modulates amygdala activity without explicit knowledge. *J Neurosci* 1998;**18**: 411–8.

33. Lau HC, Passingham RE. Unconscious activation of the cognitive control system in the human prefrontal cortex. *J Neurosci* 2007;**27**:5805–11.
34. van Gaal S, Ridderinkhof KR, van den Wildenberg VP, Lamme VA. Dissociating consciousness from inhibitory control: evidence for unconsciously triggered response inhibition in stop-signal task. *J Exp Psychol Hum* 2008;**28**:8053–62.
35. Wokke MJ, van Gaal S, Scholte HS, Ridderinkhof KR, Lamme VAF. The flexible nature of unconscious cognition. *PLoS One* 2011;**28**. https://doi.org/10.1371/journal.pone.0025729.
36. Kawakami N, Yoshhida F. Perceiving a story outside of conscious awareness: when we infer narrative attributes from subliminal sequential stimuli. *Conscious Cogn* 2015;**33**:53–66.
37. Dehaene S., Changeux JP, Naccache L, Sargent C. Conscious, preconscious, and subliminal processing: a testable taxonomy. *Trends Cogn Sci* 2006;**10**:204–11.
38. Dijksterhuis A, Strick MA. A case for thinking without consciousness. *Perspect Psychol Sci* 2016;**11**:117–32.
39. Bargh JA, Chen M, Burrows L. Automaticity of social behavior: direct effects of trait construct and stereotype activation on action. *J Pers Soc Psychol* 1996;**71**:230–44.
40. Bargh JA, Lee-Chai A, Barndollar K, Gollwitzer PM, Trötschel T. The automated will: nonconscious activation and pursuit of behavioral goals. *J Personal Soc Psychol* 2001;**81**:1014–27.
41. Bargh JA, Ferguson MJ. Beyond behaviorism: on the automaticity of higher mental processes. *Psychol Bull* 2000;**126**:925–45.
42. Papies EK, Aarts H. In: Vohs K, Baumeister R, editors. *Handbook of self-regulation: research, theory, and applications*. 3rd ed. New York: Gilford Press; 2017.
43. The reliability of a few of the very early studies has recently been called into question Bartlett T. The power of suggestion. *Chron Rev* January 30, 2013. Available from: http://chronicle.com/article/Power-of-Suggestion/136907/.
44. Sloman SA. The empirical case for two systems of reasoning. *Psychol Bull* 1996;**19**:3–33.
45. Damasio A. *Descartes' error: emotion, reason, and the human brain*. New York: Putnam; 1994.
46. Le Doux J. *The emotional brain: the mysterious underpinnings of emotional life*. New York: Simon and Schuster; 1996.
47. Smolensky P. On the proper treatment of connectionism. In: Goldman A, editor. *Readings in philosophy and cognitive science*. Cambridge (MA): MIT Press; 1993. p. 769–99.
48. Twain M. *Life on the Mississippi*. General Books LLC; 2010. p. 60–1.
49. Klinger E. Daydreaming and fantasizing: thought flow and motivation. In: Markman KD, Klein WMP, Suhr JA, editors. *Handbook of imagination and mental simulation*. New York: Psychology Press; 2008. p. 225–40.
50. Christoff K, Gordon AM, Smallwood J, Smith R, Schooler JW. Experience sampling during FMRI reveals default network and executive system contributions to mind wandering. *Proc Natl Acad Sci U. S. A* 2009;**106**:8719–24.
51. Wallas G. *The art of thought*. London: Jonathan Cape; 1926. p. 106. [It is often attributed to E. M. Foster's Aspects of Novels that came out the next year.)].
52. Tversky A, Kahneman D. Judgment under uncertainty: heuristics and biases. *Science* 1974;**185**:1124–31.
53. Kahneman D. *Maps of bounded rationality: a perspective on intuitive judgment and choice. Nobel Prize address Prize Lecture*. December 8, 2002. Available from: http://www.nobelprize.org/nobel_prizes/economic-sciences/laureates/2002/kahnemann-lecture.pdf.
54. For a particularly interesting situation that that highlights this question see Andrew Wiles. *Solving Fermat, Nova*. Available from: http://www.pbs.org/wgbh/nova/physics/andrew-wiles-fermat.html.

55. Poincaré H. *The foundations of science*. Lancaster (PA): The Science Press; 1946.
56. Salk J. *Anatomy of reality*. New York: Columbia University Press; 1983. p. 79–80.
57. Quartz D. Reason, emotion and decision-making: risk and reward computation with feeling. *Trends Cogn Sci* 2009;**13**:209–15.
58. Holton G. *Thematic Origins of scientific thought: Kepler to Einstein*. Cambridge (MA): Harvard University Press; 1973. Some of these cognitive filters are quite stable and reflect an intellectual worldview based on past experience, cultural influences, and perhaps inherent predilection. Yet it is clear that some others are only transient and instead reflect unconscious insight specific to a given situation.
59. Holton G. *The scientific imagination*. London: Cambridge University Press; 1978. p. 25–85.
60. Einstein A. Speech given on the celebration of the 60th birthday of Max Plank. *Ideas and opinions*New York: Crown Publishers; 1954. p. 226.
61. Strathern P. *Mendeleyev's dream: the quest for the elements*. New York: Berkley Trade; 2002.
62. Kekulé FA. Japp FR: Kekulé memorial lecture. *J Chem Soc* 1989;**73**:97–138.
63. Einstein credits his fantasy at the age of 16 with containing the germ of The Theory of Special Relativity Holton G. *Einstein, history and other passions*. Reading, MA: Addison-Wesley; 1996. p. 199.
64. Isenman L. Towards an understanding of intuition and its role in scientific endeavor. *Perspect Biol Med* 1997;**40**:395–403.
65. Gladwell M. *Blink: the power of thinking without thinking*. New York: Little Brown; 2005.
66. Salvi C, Bowden EM. Looking for creativity: where do we look when we look for new ideas?. *Front Psychol* 2016;**7**:161. Available from: https://doi.org/10.3389/fpsyg.2016.00161.

# CHAPTER 2
# Implicit Learning

## Contents



When you think about learning, perhaps you recall multiplication drills in grade school, conjugating irregular verbs in high school French class, or studying for final exams in college or graduate school. This willed or intentional learning, called *explicit learning*, depends strongly on conscious mechanisms. It is of secondary interest here, because intuition is grounded largely in unconscious learning and understanding.

Perhaps you instead remember realizing that you had learned something complex you did not expend any effort to master. I recall one day as a young child just knowing the address of my home. The recognition that I had learned it without trying or being aware that I was learning anything was exciting but also a little bit disquieting. How could it be, I wondered? This incidental or effortless learning is called *implicit learning*.

Something more complex that we all pick up without effort and without awareness of learning anything is the grammar of our native tongue. As young children, simply by listening to others speak we extract the underlying regularities that govern the sentence structure of our native language. We then use this knowledge below awareness to structure one grammatical sentence after another. Only later in school do we learn the rules of grammar explicitly.

Part of my goal in this chapter is to make the case that implicit learning is better than explicit learning at extracting many of the complex and often


*Understanding Intuition*
ISBN 978-0-12-814108-3
https://doi.org/10.1016/B978-0-12-814108-3.00002-6

imperfect, or noisy, patterns we encounter in the environment. It is likely an important source of information for intuition and helps ground it in scientific understanding. Implicit learning is one of a number of links that together provide an explanation for how intuition can sometimes be more knowledgeable and comprehensive than conscious intelligence.

People whose cognitive style is based strongly on intuition tend to be more aware of the role of implicit learning in mental function than others because they depend on it more directly. Arthur Reber, a pioneer in the field, coined the term "implicit learning;" his work has a prominent role in this chapter. He reveals in his book *Implicit Learning and Tacit Knowledge* that his own cognitive style is strongly intuitive:

> *I was not a particularly good "standard" student, typically bristling at directed study courses and assigned readings and papers. I found that what seemed to me to be the most satisfactory of "learnings" were those that took place through what we used to call "osmosis," that is, one simply steeped oneself in the material, often in an uncontrolled fashion, and allowed understanding to emerge magically over time. The kind of knowledge that seemed to result was often not easily articulated, and most interesting, the process itself seemed to occur in the absence of efforts to learn what was, in fact, learned.*[1]

Even babies are capable of implicit learning. In fact, it is the only kind of learning available to them because they cannot form explicit or conscious memories. Likewise, people with anterograde amnesia, who cannot form new explicit memories, can learn implicitly. Implicit learning depends on another kind of memory, which is called *implicit memory*.

Implicit memory is responsible for *priming*, or the ability of one stimulus to enhance the response to a subsequent one. For example, if patients with anterograde amnesia are given a specific word such as *boot*, they quickly forget it. However if a bit later they are asked to complete the word stem *bo–*, they are much more likely to come up with the word *boot* than another word that might fit, such as *book* or *boat*. Using their priming capacity, they can be taught to perform complex perceptuomotor tasks such as computer data entry, even though they will have no explicit memory of these skills.[2] They can also perform many of the implicit learning tasks I will describe.

Implicit learning allows us to record without effort or intent the patterns of co-occurrence and other kinds of regularity that characterize the environment. We do this not only without conscious awareness of doing so, but often without apparent motivation or reward. Chapter 4 explores the likely basis of this kind of associative learning.

The capacity of the unconscious mind to record information about the environment and recognize its inherent patterns of regularity is part of the human genetic heritage. By demonstrating that just like lower animals humans often learn without awareness of what we are learning or even *that* we are learning anything, implicit learning places intuition within an evolutionary framework. Implicit learning skills, like our other abilities, have been selected by evolution over the eons because they enhance survival. Part of their power is that they ensure that we have a wide range of knowledge of the deeper underlying regularities that characterize the environment. In turn, this can help us cope with a variety of unanticipated situations.

I will present several different approaches to studying implicit learning because it is important to ground the elusive concept of intuition in scientific evidence. I was surprised and delighted when I first read about these experiments many years ago. I continue to marvel at them.

## FREQUENCY LEARNING

In the late 1970s and 1980s, Hasher and Zacks showed that humans are "genetically prepared" to record into memory, without effort or awareness of doing so, certain basic aspects of what they referred to as the "flow of information" from the environment.[3] These include the frequency of occurrence of events, their temporal ordering, and the spatial orientation of objects. We are sensitive to these basic aspects of experience without consciously intending to be so.

In a deceptively simple study, the researchers showed subjects over 70 slides with pictures of different objects or words, telling them beforehand only that the experiment had to do with memory.[3] Each slide appeared one, two, three, or four times during the course of the presentation. The subjects were then shown some of the slides again and asked how frequently each had appeared. Because they had not paid attention to this, they thought they could not possibly answer. Nonetheless, to their surprise, the information was available to them, although their accuracy decreased for the slides shown most frequently.

Kindergarten children, schoolchildren, college students, and senior citizens all performed more or less the same. The researchers argued that the ability to record frequency develops early in life, does not improve with age as many other skills do, and does not decline significantly in old age.[4] Likewise, IQ, prior knowledge, educational background, and motivation had no influence on the ability to record frequency accurately.

However these factors did affect performance on another, more traditional memory task that instead asked subjects to remember words on a list.[5]

Letting subjects know beforehand that they would be required to remember how many times each slide appeared did not improve performance. Neither did practice sessions that included feedback. In contrast, forewarning, practice, and feedback all improved the kind of explicit memory that is required to remember words on a list.

This relative consistency across the population and life cycle, as well as independence from conscious learning aids, suggests that humans are genetically prepared to record the frequency of occurrence. It is an automatic aspect of learning that is independent of conscious effort, unlike object recall. Implicit learning skills in general tend to demonstrate a more limited range within the population than is true for general measures of intelligence such as IQ. Reber argues that this restricted variability indicates that implicit skills are evolutionarily older than those that depend more on consciousness and voluntary control.[1] By the time *Homo sapiens* appeared, the basic mechanisms underlying these capacities were standard equipment for evolving higher organisms. Because implicit learning abilities have provided less selective advantage than more recently evolved capabilities, they display less variability in the human population.

Although implicit learning skills are highly adaptive, sometimes they can have a downside. Hasher and colleagues showed that one consequence of frequency recording is that the more often we hear something, the more likely we are to give it credence.[6] For example, the first time we hear that a certain neighborhood is safe, or that a certain school system is good, we may view this as just the opinion of the particular person we heard it from. However, if we hear the same thing from several others, we are likely to accept it as true and make it part of our model of our world. Most of us would probably make a decision about where to buy a house on the basis of such repeated hearsay without feeling the need to check the statistics or ask an expert. This phenomenon also accounts for the power of propaganda, and, alas, now fake news. The rest of the chapter considers some more advanced implicit learning skills.

## ARTIFICIAL GRAMMAR LEARNING EXPERIMENTS

Arthur Reber began using artificial grammars to study implicit learning in the mid-1960s.[1] His work suggests that people have the capacity to learn the underlying regularities inherent in a complex grammatical system

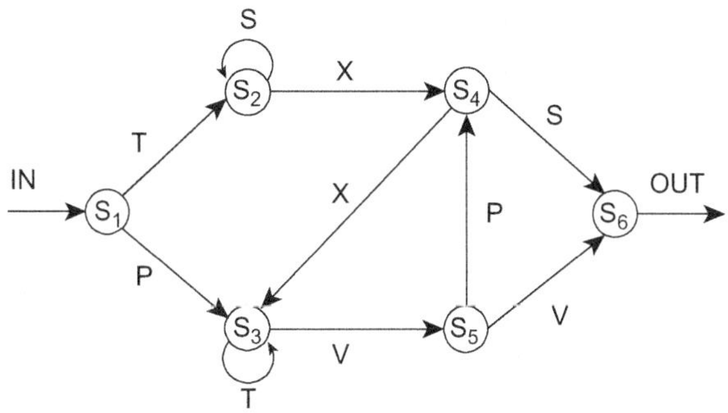


1. TXS
2. TSXS
3. TSSXXVV
4. TSXXTVPS
5. PVV
6. PTTVPS
7. PTVPXVPS
8. PTVPXVPS

**Figure 2.1** ***Example of an artificial grammar and some sample strings.*** To navigate the grammar, start at "IN" on the left and follow the arrows, sometimes choosing one path and sometimes another, to get to "OUT." Any pathway through the grammar generates grammatical strings. *(Reprinted from Reber A.* Implicit learning and tacit knowledge. *Oxford: 1993. p. 28 with permission of Oxford University Press.)*

simply by memorizing a list of examples generated by the grammar. This means that even as subjects learn one thing at an explicit level, they are actually learning something much more complex at an implicit level. His approach has been widely employed by others, and the type of artificial grammar he used is now often called a "Reber grammar."

Fig. 2.1 shows an early "Reber grammar." To navigate the grammar, one starts at "IN" on the left and follows the arrows, sometimes choosing one path and sometimes another, to get to "OUT." In artificial grammars, just as in the grammar of English or any other natural language, an infinite variety of sentences or "strings" are theoretically possible. There are alternative paths at every juncture and internal loops can be endlessly repeated. Reber realized that the potential complexity and richness of artificial grammars meant it was unlikely that subjects would be able to figure out the rules at an explicit level after learning a list of exemplars. This made it particularly well-suited for studies of unconscious learning.

In a typical experiment, subjects memorize 20 strings of up to eight letters, such as those shown in Table 2.1 on the left. Only after they have memorized the list are they told that the strings were constructed according to an underlying set of rules. Then they are given a list of new strings, such as those shown in Table 2.1 in the center and right, some of which conform

**Table 2.1** Typical learning and testing strings used in artificial grammar experiment

| Learning stimuli | Testing stimuli | |
|---|---|---|
| 1. PVPXVPS | *1. PTTTVPVS | *26. SVPXTVV |
| 2. TSSXXVPS | *2. PVTVV | 27. PVPXTTVV |
| 3. TSXS | *3. TSSXXVSS | 28. PTTVPXVV |
| 4. PVV | *4. TTVV | 29. TSXXTVPS |
| 5. TSSSXXVV | 5. PTTTTVPS | 30. TXXTVV |
| 6. PTVPXVV | 6. PVV | 31. TSSSSXS |
| 7. TXXVPXVV | *7. PTTPS | *32. TSXXPV |
| 8. PTTVV | 8. TXXTTVPS | 33. TPVV |
| 9. TSXXTVPS | 9. TSXXTTVV | *34. TXPV |
| 10. TXXTVPS | *10. PVXPVXPX | *35. TPTXS |
| 11. PTVPS | *11. XXSVT | 36. PVPXTVPS |
| 12. TSX | 12. TSSXXTVV | *37. PTVPXVSP |
| 13. TSXXTVV | 13. TXS | 38. PVPXVV |
| 14. PVPXTVPS | *14. TXXVX | 39. PTVPXVPS |
| 15. TXXTTTVV | *15. PTTTVT | *40. SXXVPS |
| 16. PTTTVPS | 16. TSXXVPS | 41. TXXVV |
| 17. TSSSXS | 17. PTTTVV | *42. PVTTVV |
| 18. TSSXXVV | *18. TXV | 43. TSSXXVPS |
| 19. PVPXVV | 19. PTTVPS | *44. PTVVVV |
| 20. TXTVPS | 20. TXXTTVV | *45. VSTXVVS |
| | *21. PSXS | 46. TSXXVV |
| | *22. PTVPPPS | *47. TXXTVPT |
| | 23. PTTTTTVV | 48. PVPS |
| | *24. TXVPS | *49. PXPVXVTT |
| | 25. TSSXS | *50. VPXTVV |

Subjects first memorize the strings on the left, which are based on the grammar in Fig. 2.1. Then they are told that the strings conform to an underlying grammar and are asked whether the novel strings in the center or right are grammatical. The asterisks indicate ungrammatical strings.
Reprinted from Reber A. *Implicit learning and tacit knowledge*. Oxford University; 1993. p. 36 with permission of Oxford University Press.

to the structure of the underlying grammar and some not. They must judge which of the novel strings are grammatical (asterisks indicate ungrammatical strings).

People who have not studied the exemplars or who have studied random strings would have to guess. Therefore they would be correct about 50% of the time. However subjects who have memorized the exemplars, even though they cannot articulate the basis on which they make their choices and often report they feel that they are guessing, typically classify the grammatical strings correctly 60% to 70% of the time, depending on the conditions.

Reber tried telling subjects beforehand that the strings to be memorized corresponded to an underlying structure, and even encouraging them to search for the rules. Instead of helping, however, this interfered with their performance. It took them longer to memorize the exemplars, they were less successful at judging which of the novel strings were grammatical, and they often came to erroneous conclusions about the underlying regularities. At least in situations in which the structural regularities are highly complex, consciously searching for them is more of a hindrance than a help. Or, as Reber straightforwardly concludes, "*Looking for rules won't work if you cannot find them*."[1(p48)]

Yet under certain other conditions, encouraging subjects to search for the rules does increase performance: for example, when the letter strings are longer and therefore give more information about the loops, or when the exemplars are ordered according to their underlying form so as to enhance the salience of the rules. Reber says of these situations, "*Looking for rules will work if you can find them*."[1(p49)] It is difficult to argue with the simple sense of Reber's conclusions, yet they have surprisingly profound implications. They suggest that the unconscious mind may be better equipped than the conscious mind to deal with some difficult learning situations.

## What the Brain Stores

Reber argues that the artificial grammar studies show that the unconscious is able to extract from the memorized strings the underlying structural regularities that unite them. Subjects establish abstract or "deep" representations of the pattern of covariance between the letters of the strings memorized, and on this basis they judge whether the novel strings are grammatical.[7,8] He feels that the trace of this information that the brain stores relates in some manner to the abstract rules of the grammar. Yet he points out that this does not mean it directly corresponds to the rules of the grammar. Moreover other grammars than the one the experimenter used to generate the strings of letters also may account for the same structural relationships between the letters.[1]

These experiments are considered classics in the field of psychology, yet it is not always clear what information subjects use to judge whether novel strings are grammatical. Some researchers have argued that the brain encodes only the different examples memorized and not the regularities inherent in the relationships between the letters.[9] Subjects then use the similarity of the exemplars to the novel strings to judge whether the novel strings are grammatical. Another group argues that they use only the presence of particular two- or three-letter strings, called *bigrams* or *trigrams*, similar to those in the exemplars to judge novel strings as gramatical—much as the conscious mind might approach the task.[10] A more recent study shows that both of these additional factors can influence judgments under certain conditions. However subjects can still judge grammaticality at better than chance levels even when the novel strings are carefully designed so that both types of cues are constant in the both grammatical and ungrammatical strings.[11]

## Experiments With Babies

Additional support for Reber's view comes from experiments with babies.[12] Infants generally stare longer when they are exposed to unfamiliar as opposed to familiar stimuli. This provides a way to measure whether they can recognize given regularities at an implicit level. In one study, 7-month-old babies heard 2 min of three-syllable sentences, or strings, with the form ABB, such as "ga ti ti" or "li na na." They then heard 12 individual strings, some of the form ABB, such as "we fe fe," and others of the form ABA, such as "wo fe wo." Strings with the unfamiliar grammatical ABA form triggered longer staring. This shows that 7-month-olds are capable of learning grammatical structure and that they can apply or *transfer* this knowledge to a novel series of nonsense syllables.

Researchers have also demonstrated transfer of knowledge in adult grammar learning experiments. Subjects who are tested with strings composed of a different set of letters than the exemplars they memorized earlier can still determine whether the novel strings conform to the *structure* of the grammar—that is, preserve its internal relationships. Transfer can occur even between sensory modalities, such as between tones and letters.[13] One delightful early multimodal experiment showed transfer between the letters of the grammar and sequences of a squeak, drip, hiss, clang, and flash.[14]

The researchers who support the view that the brain encodes only the exemplars that are memorized, and not their deeper underlying regularities, have an alternate explanation for the transfer results. Subjects generate the required abstractions on the spot in the process of comparing the novel string with the exemplars.[15,16] Thus for each novel string, they run through the list of memorized exemplars looking for a match of relational characteristics. Yet this cannot account for the transfer results with babies. They cannot form explicit memories of the exemplars nor can they perform the kind of on-the-spot comparison this would require.

## SEQUENCE LEARNING EXPERIMENTS

In another approach used to investigate implicit learning, called *sequence learning*, subjects learn temporal patterns of events little by little during the course of the experiment. The design of most sequence learning experiments is broadly similar. A signal such as a flashing light appears in one of four or more positions at the bottom of a computer screen, or perhaps in one of its four quadrants. The subject simply presses a specific key on the keyboard designated to correspond to the position of the signal on the screen. A computer measures the subject's reaction time (the time it takes to press the key) in 0.001-s intervals.

In the simplest sequence learning experiments, the signal moves from position to position in a fixed sequence. In one experiment, subjects watched a star appear at the bottom of the screen at one of four equally spaced positions. If A, B, C, and D stand for the four respective positions of the star, it appeared on the screen according to the sequence DBCACBDCBA. The sequence was repeated continuously 100 times without a break. Over the trial, subjects took less and less time to respond to the signal.[17]

In order to be sure that the decrease in reaction time did not result from some incidental aspect of the task, at the end the computer switched the star to another sequence or random pattern of movement. (For example, the decrease could instead reflect the fingers warming up and becoming nimbler with practice.) Since reaction time increased sharply with the change, this showed that the subjects had learned the old sequence during the earlier trials.

The decrease in reaction time could indicate either implicit or explicit learning. To determine the extent of *explicit* learning, the researchers first asked subjects whether they were aware of a sequence and what they could remember about it. Then they asked them to generate it one step at a time on the computer screen by pressing the correct key for each position. The subjects were required to come up with the right answer for each step, even if this meant making a series of sequential guesses. This *generation test* is considered a less demanding test of conscious knowledge than reporting about the sequence verbally, because unconscious knowledge is more likely to inform the sequence they tap on the computer.

Some subjects showed explicit knowledge of the entire sequence, and some, part of it. About a third showed no evidence of explicit learning. Nonetheless their average reaction time decreased by almost a full second over the course of training, which indicated implicit learning of the sequence. Subjects with antegrade amnesia, who could not form new conscious memories, also showed faster reaction times with training. For them, all learning was necessarily implicit.

A more recent sequence learning experiment reported in an article entitled "On the Benefits of Not Trying"[18] demonstrated for sequence learning what Reber found for artificial grammar learning: when the sequences are complex, looking for them interferes with learning them at an implicit level. In some trials, subjects were told to respond as quickly (and as accurately) as they could, a condition that favors implicit learning. In others, they were told to look for a repeating sequence, a condition that favors explicit learning. With a simple repeating sequence, reaction times decreased comparably under both conditions. All subjects were aware that there was a sequence and were able to remember some successive elements. However, under explicit learning conditions, they remembered twice as much about the sequence.

A more complex sequence was used in other trials that were interspersed with the simple sequences trials. It had a random number between each element of the sequence and a different one for each repeat (Fig. 2.2). When subjects looked for this more difficult, or noisy, sequence as they trained (the condition favoring explicit learning), no decrease occurred in reaction time. But they were aware afterward that there was a sequence and could reproduce a few elements. They could do neither when they responded as quickly as possible as they trained. Yet reaction time decreased, indicating that implicit learning had occurred.

| EASY SEQUENCE | | DIFFICULT SEQUENCE |
|---|---|---|
| 1 3 4 1 2 4 2 3 1 4 | 1st repeat | 1 3 4 1 2 4 2 3 1 4 |
| 1 3 4 1 2 4 2 3 1 4 | 2nd repeat | 1 2 4 3 2 1 2 4 1 2 |

**Figure 2.2** ***Testing implicit and explicit learning with hard versus easy sequences.*** Subjects trained with both a simple repeating sequence and a more difficult sequence in which every other element was random, under two different conditions. In one condition, they responded as quickly as they could, a condition that favors implicit learning. In the other, they tried to find the sequence, which favors explicit learning. *(From Fletcher P, et al. On the benefits of not trying: brain activity and connectivity reflecting the interactions of explicit and implicit sequence learning.* Cereb Cortex *2005;7:1002–15. Reprinted with permission by Oxford University Press.)*

Brain imaging showed that implicit learning of the complex sequence was accompanied by specific time-dependent alterations in brain activity. These changes were attenuated with explicit learning, thereby documenting at yet another level that consciously trying to find a complex pattern can interfere with implicit learning of the pattern. The notion that trying harder in complex learning situations is not necessarily the best strategy will probably not surprise those who favor Reber's "osmosis" as a learning style.

The complexity of the patterns that researchers have shown we are able to learn implicitly is astounding. In some sequence-learning experiments the pattern is so complex and obscure that the likelihood that anyone will figure out the sequence is nil. Moreover misleading instructions may be given by the researchers to obscure the objectives of the test. Often subjects are completely unaware of what they are learning and, moreover, unaware that any learning is occurring.

In one such study, the signal appeared successively in one quadrant or another of the computer screen in blocks of seven.[19] The researchers were interested only in the response to the seventh signal, the last in each block. Where the seventh signal appeared was determined by the quadrant in which the first, third, fourth, and sixth signals of that block had already appeared. The position of the second and fifth signals in each block was always completely random.

There are a total of 24 possible configurations for the first, third, fourth and sixth signals. Six of them were arbitrarily assigned to correspond to one of the four quadrants for the seventh signal, six to another, etc. Table 2.2 summarizes the rule used by the computer to generate the seventh signal. The rule is challenging for the conscious mind to comprehend, to say nothing of the difficulty of picking it up one element at a time, as the unconscious mind must during the course of the experiment. To make the task even more difficult, for the seventh trial in each block the screen was filled with distracting figures, and the signal overwrote one of them. During the experiment, subjects had to learn implicitly which trials in the block were the important ones, as well as the different configurations of the first, third, fourth, and sixth signals that determined in which quadrant the seventh signal would go.

Needless to say, learning was not instantaneous. Training extended over 6 days and required subjects to respond to about 4600 sets of seven signals. During this period, the average reaction time to the seventh signal decreased

**Table 2.2** Acquiring very complex implicit learning

| Trial | | | | | | |
|---|---|---|---|---|---|---|
| **1** | **2** | **3** | **4** | **5** | **6** | **7** |
| 3 | R | 1 | 4 | R | 2 | 1 |
| 3 | R | 4 | 2 | R | 1 | 1 |
| 2 | R | 4 | 3 | R | 1 | 1 |
| 4 | R | 1 | 2 | R | 3 | 1 |
| 1 | R | 2 | 4 | R | 3 | 1 |
| 2 | R | 3 | 1 | R | 4 | 1 |
| 2 | R | 4 | 1 | R | 3 | 2 |
| 1 | R | 3 | 4 | R | 2 | 2 |
| 1 | R | 2 | 3 | R | 4 | 2 |
| 4 | R | 3 | 1 | R | 2 | 2 |
| 4 | R | 2 | 3 | R | 1 | 2 |
| 3 | R | 1 | 2 | R | 4 | 2 |
| 4 | R | 1 | 3 | R | 2 | 3 |
| 1 | R | 3 | 2 | R | 4 | 3 |
| 4 | R | 2 | 1 | R | 3 | 3 |
| 3 | R | 2 | 1 | R | 4 | 3 |
| 2 | R | 3 | 4 | R | 1 | 3 |
| 1 | R | 4 | 3 | R | 2 | 3 |
| 3 | R | 2 | 4 | R | 1 | 4 |
| 3 | R | 4 | 1 | R | 2 | 4 |
| 1 | R | 4 | 2 | R | 3 | 4 |
| 2 | R | 1 | 3 | R | 4 | 4 |
| 2 | R | 1 | 4 | R | 3 | 4 |
| 4 | R | 3 | 2 | R | 1 | 4 |

Where the seventh signal appeared in each block was determined by the quadrant in which the first, third, fourth, and sixth signals of the block had already appeared. The position of the second and fifth signals was always random. Of the 24 possible configurations for the first, third, fourth, and sixth signals, six were arbitrarily assigned to correspond to one of the four quadrants for the seventh signal, six to another, etc.

Reprinted from Lewicki P, Czyzewska M, Hoffman H. Unconscious acquisition of complex procedural knowledge. *J Exp Psychol Learn Mem Cogn* 1987;**13**:523–30.

by about one-half (Fig. 2.3). In the last session, the rules were changed so that the seventh signal appeared in the quadrant opposite where it had been before. In response, the reaction time increased about 0.1 s, as shown by the arrow. (It did not go back to what it had been in the beginning, because the subjects' knowledge of which trials in each block were the important ones, as well as which six configurations of these trials were equivalent to each other, was still relevant.[20])

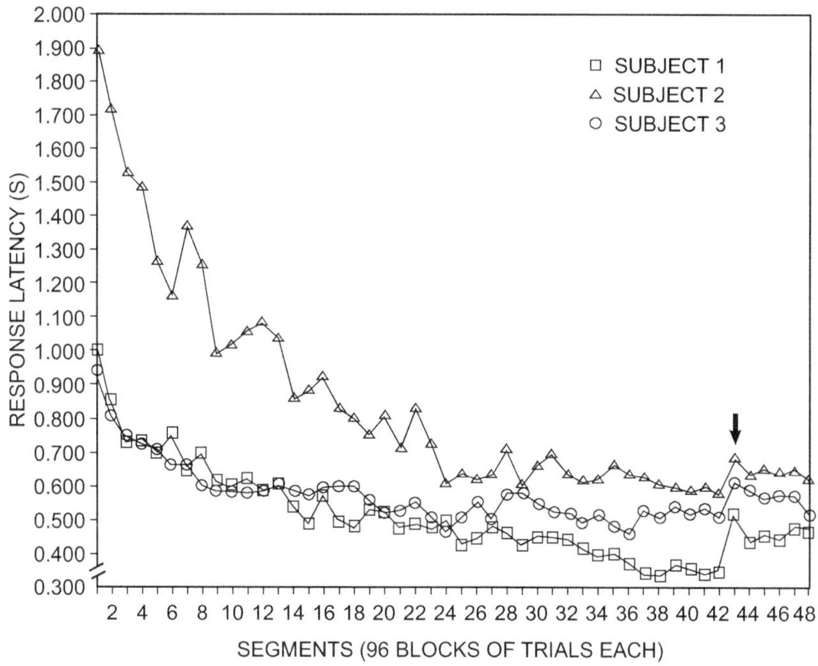


**Figure 2.3** ***Decrease in reaction time throughout the training period with very complex implicit learning.*** At the arrow, the signal appeared in the quadrant opposite where it had been before and the reaction time increased slightly, showing that implicit learning of the grammar had occurred. *(From Lewicki P, Czyzewska M, Hoffman H. Unconscious acquisition of complex procedural knowledge.* J Exp Psychol Learn Mem Cogn *1987;**13**:523–30.)*

Unsurprisingly, at the end of the experiment the subjects had no explicit knowledge of the rules; moreover they were unaware that the appearance of the seventh signal followed a rule. The patterns in this experiment are so complex that they seem far removed from any we might encounter in the environment. Yet with repeated exposure, our genetic endowment allows us to capture below awareness even regularities that are as complicated as those in this study.

Another experiment, by Cleeremans and McClelland, combined aspects of artificial grammar learning and sequence learning.[21] By definition, the number of different paths through a grammar is unlimited. The researchers started with the artificial grammar shown in Fig. 2.4. Then they used a computer to generate a nonrepeating series of sequences, or "strings," that determined where on a screen the signal would appear. The grammar used six

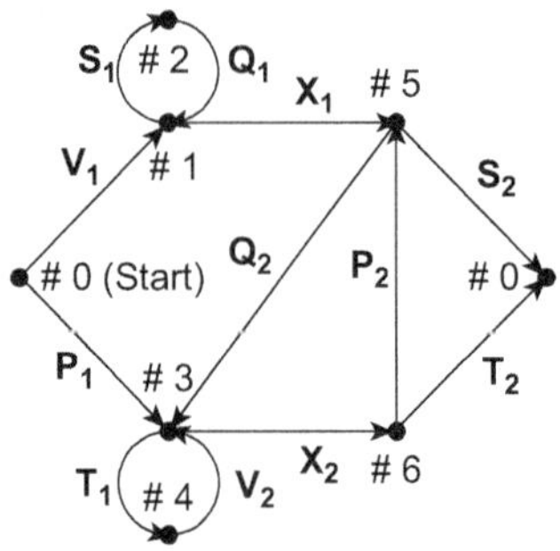


**Figure 2.4** ***Combining artificial grammar learning and sequence learning.*** An artificial grammar was used by a computer to generate a series of ever-changing strings that determined where on a screen the signal would appear. The computer also changed 15% of the positions in the sequences at random. *(From Cleeremans A, McClelland JL. Learning the structure of event sequences.* J Exp Psychol Gen *1991;***120***:235–53.)*

letters, each of which labeled two of its arcs (lines connecting two nodes). To make learning even more difficult, the computer changed 15% of the positions in the sequences at random.

These sequences are statistical rather than fixed or deterministic. They depend on an underlying grammar with a number of choice points, each letter is used twice in the grammar, and some of the information is altered at random. (Some of the data was random in the sequence learning experiments described previously, but it always came in the same position in the trial.)

Fig. 2.5 shows the decrease in reaction time for subjects for the grammatical as well as the random, or ungrammatical, sequence elements over the 20 training sessions. Learning occurred slowly, which allowed the researchers to track its progression. Subjects first picked up which positions were more likely to follow each other, and over time they began to develop some knowledge of higher-order contingencies. The unconscious learning demanded by this task directly demonstrates our ability to extract statistical regularity from noisy samples to learn about underlying patterns.

## HOW IMPLICIT IS IMPLICIT LEARNING?

The question of just how implicit or explicit the learning is that occurs in implicit learning studies has generated an immense amount of controversy.[22–26] Until relatively recently, much of the work in the field was devoted to resolving this issue. Critics have claimed that subjects really do have explicit, or conscious, knowledge of what they have learned. However

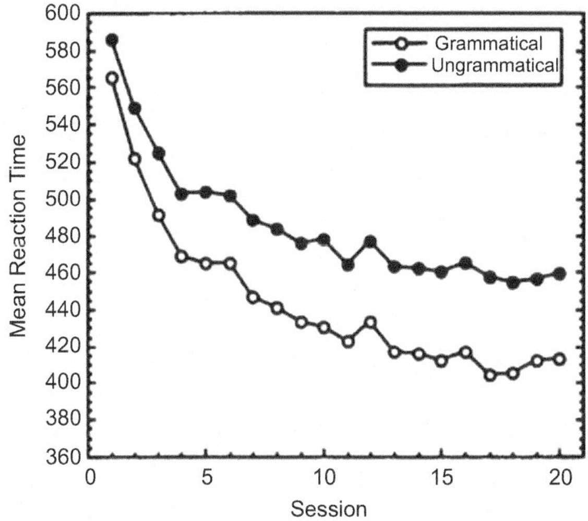


**Figure 2.5** ***Mean reaction times for grammatical and random (ungrammatical) strings in an experiment combining artificial grammar learning and sequence learning.*** Both decreased with practice, so the difference between the two traces indicates implicit learning of the underlying grammar. *(From Cleeremans A, McClelland JL. Learning the structure of event sequences.* J Exp Psychol Gen *1991;**120**:235–53.)*

they do not verbalize this knowledge when questioned, because they are hesitant to say anything if they are not sure it is right. Critics have also argued that just because subjects appear to be unaware of what is guiding their responses does not mean that the underlying knowledge is completely unconscious.

How implicit any response is judged to be depends to a large extent on the test that is used to measure explicit knowledge. Critics propose that sensitive enough tests would show that subjects really do have explicit knowledge. When given part of a novel string from an artificial grammar, for example, sometimes they can predict what the next letter would be, or they can say which part of a string makes it seem grammatical or ungrammatical. But does the ability to answer these questions mean that the knowledge is held explicitly?

It does seems to argue that some knowledge related to the grammar or the sequence can in fact be accessed consciously. However the right answer can appear in consciousness as a vague feeling or a hunch, or even a strong conviction, while the underlying information on which it is based remains inaccessible. Just as a child implicitly knows that a verb tends to follow a subject, we can know an answer without knowing why it is right. Moreover after coming up with such an answer, sometimes we can bring some of the underlying information on which it is based to awareness, or infer what it must have been.[1] But again this is different from having used this information consciously to make the judgment in the first place.

Knowing whether some implicit learning can be entirely isolated from explicit learning is an important scientific question. Yet the possibility of sophisticated intuition does not depend on the answer. Such a focus necessarily excludes some of the most intriguing questions about implicit learning as it relates to intuition. Among these are the characteristics of different degrees of implicit learning, the possible interaction among different kinds of implicit learning, as well as how implicit and explicit knowledge interact. These questions are increasingly the subject of current research.

An alternate perspective stresses the continuity of implicit and explicit learning. Reber, for example, maintains that implicit and explicit learning lie on a continuum with implicit learning primary. He argues that what is held implicitly is always more than what is held explicitly. He says of the controversy, "While it is misleading to argue that implicitly acquired knowledge is completely unconscious, it is not misleading to argue that the implicitly acquired epistemic contents of mind are always richer and more sophisticated than that which can be explicated."[1(p64)] Many experiences of intuition appear to support this view.

Reber contrasts this idea of the "primacy of the unconscious" or "the implicit stance" to what he calls the "consciousness stance." For example, in a 1994 review article, Shanks and St. John declared that the evidence for implicit learning is highly questionable and that learning in humans is in most cases linked to consciousness.[24] In a 2005 review article, Shanks concluded by saying that there is reasonable doubt about the existence of implicit learning, and in a 2010 review he expressed the same view.[25,26]

This subtle shift, or softening, reflects the increased strength of the experimental evidence for implicit learning, in part in response to questions that critics have raised. It also likely reflects the ever-growing recognition of the importance of unconscious cognition. Nonetheless, especially with respect to sophisticated cognition, conscious processes still tend to be the default assumption unless specifically shown otherwise. What Reber wrote in 1993 rings true today:

> *... from this perspective, consciousness takes priority; awareness and self-reflection become the central features of human cognitive function. The cognitive unconscious is dealt with by exclusion and implicit processes have to be defended against*

*claims of residual awareness. The popularity that this consciousness stance has in contemporary psychology is in no small way due to the very poignant sense that awareness and sense of self are what defines our very existence and, therefore, are the features that make us human. … The stance that gives the unconscious and the implicit priority is one that, one might argue, strips away something fundamentally human from our characterization of ourselves.*[1(p69)]

## EXPLORING THE RELATIONSHIP BETWEEN IMPLICIT AND EXPLICIT KNOWLEDGE

Dostbecqz and Cleeremans used a popular technique developed to separate implicit and explicit knowledge to begin to clarify their relationship to each other.[27] Called *process dissociation*, it addresses the concern that tests used to measure conscious knowledge are often influenced by unconscious knowledge, as well as the other way around.[28] Process dissociation gives estimates of both.

The researchers used a serial reaction task with a simple repeating sequence in which each element was determined by the prior two elements.[28] For one group of subjects, during training the next element in the sequence appeared 250 ms after they had responded to the previous element, as is typical in sequence learning experiments. For the other group, the next element instead appeared immediately after they had responded to the previous one, or with a 0-ms delay. For both groups, the signals were shown for the same amount of time. But the memory traces of subjects who trained with a 0-ms delay had no time to develop before the next stimulus appeared. Therefore they were weaker, or of lower quality, than the memory traces of subjects who trained with a 250-ms delay.

After training, all subjects had to generate 96 elements of the repeating sequence by themselves on the computer, guessing if they did not know. This is called the *inclusion* task because it *includes* both conscious knowledge of the sequence and unconscious knowledge, such as hunches and what the "fingers know." In contrast, in what is called *the exclusion* task, they had to produce a sequence of 96 numbers *as different as possible* from the one on which they had trained. To avoid reproducing the next sequence element, so the reasoning goes, they would have to be conscious of it because only knowledge that is conscious can be vetoed. The extent to which they nonetheless reproduced elements of the sequence therefore provided an estimate of their unconscious knowledge. Together the inclusion and exclusion results allowed estimates of the amount of implicit and explicit learning.

Both groups were able to reproduce some elements of the sequence correctly on the inclusion task. However, only those trained with the longer 250-ms interval between a response and seeing the next element were able to exclude what they had learned implicitly of the sequence, and thus had conscious knowledge of it, at least as measured by this test. Also, only those who trained with the larger interval could recognize legitimate from illegitimate sequence fragments in another test of their knowledge.

However additional experiments showed that even the knowledge of subjects who trained with a 250-ms delay was not necessarily *fully* conscious. Their ability to reproduce the sequence did not match their confidence in their answers, as it did in subjects trained in another experiment with a longer, 1500-ms delay.[29] Also, those trained with a 250-ms delay were less accurate at recognizing legitimate from illegitimate sequence fragments.

Taken together, these studies suggest that the divide between the unconscious and conscious is a continuum.[30] Successively increasing the time interval between the response to one stimulus and the presentation of the next during training leads to better-quality memory traces that give subjects more and more conscious control over their use. Cleeremans argues that this continuum implies that consciousness builds on unconscious learning mechanisms and systems of representation. Additional neural mechanisms associated with consciousness can become activated when memory strength reaches certain levels and provide the possibility of increasingly flexible control over behavior. Cleeremans and colleagues call their view of consciousness "a graded, dynamic perspective."[30] It emphasizes the graded nature of consciousness, yet it leaves room for the apparent all-or-none characteristics of some of its aspects.

They also extend the continuum to include knowledge that has become automatic through overlearning, such as knowing how to serve a tennis ball, knowing the words of the national anthem, or knowing that a cat is a cat. In contrast to either implicit or explicit knowledge, automatic knowledge is made up of memory traces so strong that the behavior associated with them is once again not subject to conscious control. Extra effort is necessary to access and then change them. Think of the effort it would take to start referring to cats as dogs, and vice versa! This distinction between overlearned memory traces and implicit ones that have not reached awareness is useful in thinking about intuition. It separates the automatic habitual knowledge that is sometimes considered intuition from the novel, previously inaccessible knowledge that I stress as its core.

Cleeremans' group works within the associative or connectionist tradition of information processing, which is modeled on the way neurons connect to each other. Connectionism overlaps strongly with what I call *intuitive processing*. It depends on parallel-interactive, or distributed, connectivity, which is discussed in Chapter 4.

Briefly, when neurons, or processing units in connectionism, fire at the same time, the connections between them are strengthened. In the future, when one fires, the others are also more likely to fire. Eventually they form networks that store knowledge of different regularities in the environment, such as knowledge of the different features of an object, or of different objects that frequently co-occur, in the strength of the connections between different neurons or units.

The information encoded in the strength of the connections between units is unavailable to awareness. Rather, it structures the network. However the pattern of activation of its output in response to different environmental stimulation is potentially available to consciousness.[30] With increased exposure, these memory traces, or representations, become stronger, which means that they involve more neurons or processing units and do so more intensely. If they become strong enough, they might make their way to awareness.[31]

The output patterns can also become the input for other networks.[32] With sequence learning experiments, for example, the output feeds into motor networks that prepare and control the finger motions. The higher-level networks can work with weak memory traces, and they update continuously.[33]

Cleeremans stresses that within the connectionist framework, even weakly encoded, poor-quality traces are active. All encounters with the world, no matter how insignificant, change the brain.[32] In addition to any automatic influence they might have on behavior, weak traces can influence consciousness from below awareness. Via their association with other, stronger traces below awareness, they can slightly nudge consciousness in a certain direction, or suggest subtle shadings of meaning.[30]

This view of the importance of memory strength also begins to evoke a model to explain how a web of interactive implicit knowledge in which each strand is too weak on its own to appear in awareness, might do so in association if together they achieve an important need or goal. Weakly encoded representations might also join below awareness with potentially conscious information that is nonetheless tacit. One reason that it might remain tacit is that it is irrelevant to ongoing conscious priorities.[34] Another possibility is that it might be an unattended higher-level meaning of contents available to consciousness,[35] which again seems unimportant to current priorities. The addition of different pieces of weakly encoded information might then expand, refine, or recontextualize the potentially conscious information. This might eventually result in the growth of a full-blown intuition below awareness, perhaps responsive to a hidden goal, or one previously relegated to the back burner.

## IMPLICIT LEARNING AND THE EVALUATIVE ASPECT OF INTUITION

Studies by Cleeremans and colleagues, as well by other researchers, have begun to chart some of the murky terrain out of which intuition arises. They argue that knowledge can have many gradations between being deeply unconscious and fully conscious. Ongoing work continues to excavate this fertile intermediate terrain between the unconscious and conscious aspects of mind.

Some researchers who have repeated Cleeremans' studies with a new twist or two find that subjects can *exclude* learned sequence elements even when they train with a 0-ms delay, and thus have poor-quality memory traces. In accordance with the rationale behind the process dissociation technique, one such study concluded that all the learning that occurred was conscious.[36] However subjects were promised money for good performance in this experiment.

A subsequent experiment by Fu, Fu, and Dienes demonstrated that the potential for reward can increase the ability of subjects to exclude knowledge even when the memory traces are weak.[37] First, the researchers trained subjects with a 0-ms delay between sequence elements. Then, they promised half of them a small sum of money for successful performance. They found that all were able to exclude (i.e., not reproduce) fragments of the training sequence to some degree. However the group that was promised money was more successful at the exclusion task, which suggests that enhanced motivation can actually *increase the influence of previously formed memory traces.*

Neither the ability of subjects who train with 0-ms delay to exclude the influence of previously formed memory traces, nor the ability of enhanced motivation to increase the influence of these traces need necessarily mean that the traces come to awareness. Norman and coworkers used an even more challenging task to explore the possibility that exclusion could occur without the sequence coming to consciousness. During training, the signal, in addition to changing position according to a repeating sequence, also changed color and shape randomly.[38] Moreover the exclusion condition required subjects to give, not the next element in a sequence fragment, but instead a position *rotated* either one or two positions clockwise or counter-clockwise, from the next element, as they were directed trial by trial. Again, subjects who trained with a 0-ms delay showed successful exclusion. This was true even for some who could not verbalize that position, rather than color or shape, was the critical variable. Thus they could not have conscious knowledge of the sequence.

The assumption central to the process dissociation technique that information needs to come to consciousness before it can be vetoed does not always hold. The possibility of having flexible control over implicit knowledge intersects here with the evaluative side of intuition. According to connectionism, as I mentioned before, even when memory traces, or output patterns of activation across networks, are not strong enough to come to awareness, they can feed into other networks.[32,33] Moreover the higher-level networks can begin to work with weak traces. Some of these higher-level networks seem to report evaluative, or *metacognitive* information, the online evaluative information that guides behavior as it occurs.[39–42] These signals tend to appear at a conscious level as a sense of rightness wrongness, or familiarity. However, like the underlying knowledge itself, the evaluative signals can be graded.[42–45] Even when they do not come to awareness, they can influence judgment and decision making.

Evaluative feelings have an important role in intuition. They can guide executive functions even when the underlying knowledge on which they report remains implicit and even when they themselves remain hidden.[45,46] The findings from implicit learning studies join results from several other realms to suggest that flexible executive functioning sometimes may occur below consciousness. I will mention several examples here, and more are provided in Chapter 8.

Under hypnotic suggestion, subjects can follow the instruction to omit a number, say four, when counting from one to five, without having an awareness of vetoing it.[47] Moreover, subliminal stimuli can initiate one of two previously set tasks that evoke activity in different regions of the brain: i.e., determine whether a word refers to a concrete object, or alternatively, whether it is bisyllabic.[48] Chapter 8 explores different kinds of evaluative feelings as well as their potential interaction with unconscious executive function in greater depth.

## HOW IMPORTANT IS IMPLICIT LEARNING TO BREAKTHROUGH INTUITION?

To really grapple with the potential role of implicit learning in sophisticated intuition, we have to revisit the finding that implicit learning shows little variation in the population as a whole. This would mean that the next Einstein, you, and I all have roughly comparable implicit learning abilities. However, by definition, the next Einstein will have intuitions of much greater profundity.

Some researchers have found interesting differences in implicit learning ability among people. Kaufman, for example, showed that subjects who test high on measures of openness to experience tend to do better on tests of implicit learning.[4] But the variability among individuals in implicit learning is much less than it is for explicit learning. It is possible that small differences in implicit learning ability confer great advantages. However other factors, either in addition to or instead of a slightly greater than normal capacity for implicit learning, may also account for why some people are capable of intuitions of astounding depth.

Unintentional or *incidental* implicit learning, the topic of most of the chapter, provides one source of knowledge for intuition. It ensures that everyone has a range of information about the environment, some of which may prove useful in the future. The experiment mentioned in the last section in which subjects after they trained were offered money for successful performance on an exclusion task shows that current motivation can increase the influence of previously encoded traces.[37] When we need it, implicit learning may sometimes just appear in awareness.

Everyone is also likely to record below awareness patterns of regularity that hold special interest for them. High motivation can greatly increase the rate of at least certain kinds of implicit learning, along with its appearance in awareness.[46] Intuition may depend as much on motivated as it does unmotivated implicit learning.

A few studies have confirmed that implicit learning can be influenced by past knowledge,[49–51] although much more work is needed in this area. Several have shown that expectations about the data may influence the extent of implicit learning.[49,50,52] In a similar vein, whether information is recorded at all can depend on distinctions acquired in the past. For example, only subjects who are well-versed in atonal music can learn at an implicit level the transformations that provide the melodic structure, or grammar, of specific pieces of atonal music.[51] Taken together, this link to motivation and to past knowledge acquisition suggests that something we learn today could influence the patterns our unconscious mind finds worthy of recording at an implicit level tomorrow. This, in turn, implies that those with greater, curiosity, knowledge, and intelligence would record more complex patterns at an implicit level.

Nevertheless, even the potential for motivated implicit learning to bootstrap on conscious learning does not seem to tell the whole story about individual differences in the depth of intuitions. The unconscious mind on its own seems to have the ability to connect different strands of tacit knowledge in novel and productive ways if together they support underlying needs. Some of the vignettes in this book, such as the "blink of the eye" encounter, attest to this ability of the unconscious mind.

Intuition seems to draw on a variety of different sources of information that mix together below awareness to fulfill its aims; implicit learning is likely an important one. However we know only a little about how mixing below awareness might happen. We do know that the ability of the mind to join information in complex, unexpected ways, depends on something that varies more widely in the population than implicit learning. Creativity, which is central to higher-level intuition, is highly inheritable. Genetic makeup apparently predisposes some brains to become more functionally interconnected than others and therefore to have the ability to connect more diverse pieces of information and realms of knowledge in productive ways. Many different factors are likely to make this possible.[53] The next chapter will focus on the kind of parallel interconnected information processing that is crucial to intuition as well as creativity.

## REFERENCES


1. Reber A. *Implicit learning and tacit knowledge: an essay on the cognitive unconscious*. Oxford: Oxford University Press; 1993. p. 22.
2. Glisky EL, Schacter DL, Tulving E. Computer learning by memory-impaired patients: acquisition and retention of complex knowledge. *Neuropsychologia* 1986;**24**:313–28.
3. Hasher L, Zacks RT. Automatic and effortful processes in memory. *J Exp Psychol* 1979;**103**:356–88.
4. For a more recent study characterizing some differences in implicit learning abilities among different groups see, for example, Kaufman SB, DeYoung CG, Gray JR, et al. Implicit learning as an ability. *Cogn* 2010;**116**:321–40.
5. Zacks RT, Hasher L, Sanft H. Automatic encoding of frequency: further findings. *J Exp Psych Learn Mem Cogn* 1982;**8**:106–16.
6. Hasher L, Goldstein D, Toppino T. Frequency and the occurrence of referential validity. *J Verb Learn Verbal Beh* 1977;**16**:107–12.
7. Reber A, Allen R. Analogic and abstraction strategies in synthetic grammar learning. *Cogn* 1978;**6**:189–221.
8. McAndrews MP, Moscovitch M. Rule-based and exemplar-based classification in artificial grammar learning. *Mem Cogn* 1985;**13**:469–75.

9. Brooks LR, Volkey JR. Abstract analogies and abstracted grammar. *J Exp Psychol Gen* 1991;**120**:316–23.
10. Perruchet P, Pacteau C. Synthetic grammar learning: implicit rule abstraction or explicit fragmentary knowledge? *J Exp Psychol Gen* 1990;**119**:264–75.
11. Lieberman MD, Chang GY, Chiao J, et al. An event-related fMRI study of artificial grammar learning in a balanced chunk strength design. *J Cogn Neurosci* 2004;**16**:427–38.
12. Marcus GF, Vijayan S, Bandi Rao S, et al. Rule learning by seven-month-old infants. *Science* 1999;**283**:77–80.
13. Altmann GTM, Dienes D, Goode A. Modality independence of implicitly learned grammatical knowledge. *J Exp Psychol Learn Mem Cogn* 2005;**21**:899–912.
14. Manza L, Reber AS. Representing artificial grammars: transfer across stimulus form and modalities. In: Berry DC, editor. *How implicit is implicit learning?*. Oxford: Oxford University Press; 1998.
15. Brooks LR, Volkey JR. Abstract analogies and abstracted grammar: comments on Reber (1986) and Mathews et al. (1989). *J Exp Psychol Gen* 1981;**120**:316–23.
16. Redington M, Charter N. Transfer in artificial grammar learning: a reevaluation. *J Exp Psychol Gen* 1996;**125**:123–38.
17. Willingham DB, Neissen MJB. On the development of procedural knowledge. *J Exp Psychol Learn Mem Cog* 1989;**14**:1047–60.
18. Fletcher PC, Zafiris O, Frith CD, et al. On the benefits of not trying: brain activity and connectivity reflecting the interactions of explicit and implicit sequence learning. *Cereb Cortex* 2005;**7**:1002–15.
19. Lewicki P, Czyzewska M, Hoffman H. Unconscious acquisition of complex procedural knowledge. *J Exp Psychol Learn Mem Cogn* 1987;**13**:523–30.
20. For a study showing that the decrease in reaction time came primarily from learning to anticipate where to look for the next signal see Sandler M. On learning complex procedural knowledge. *J Exp Psychol Learn Mem Cogn* 1989;**15**:1061–9.
21. Cleeremans A, McClelland JL. Learning the structure of event sequences. *J Exp Psychol Gen* 1991;**120**:235–53.
22. Dulany DE, Carlson RA, Dewey GI. A case of syntactical learning and judgment: how conscious and how abstract? *J Exp Psychol Gen* 1986;**113**:541–55.
23. Perruchet D, Vinter A. The self-organizing consciousness. *Behav Brain Sci* 2002;**25**: 297–330.
24. Shanks DR, St John MF. Characteristics of dissociable learning systems. *Behav Brain Sci* 1994;**17**:367–447.
25. Shanks DR. Implicit learning. In: Lamberts K, Goldstein RL, editors. *Handbook of implicit learning*. London: Sage; 2005. p. 202–22.
26. Shanks D. Learning: from association to cognition. *Annu Rev Psychol* 2010;**61**(2010): 273–301.
27. Destrebecqz A, Cleeremans A. Can sequence learning be implicit? *Psychon Bull Rev* 2001;**8**:343–50.
28. Jacoby LL. A process-dissociation framework: separating automatic from intentional uses of memory. *J Mem Lang* 1991;**30**:513–41.
29. Destrebecqz A, Cleeremans A. Temporal effects in sequence learning. In: Jimenéz J, editor. *Attention and implicit learning*. Amsterdam: John Benjamins; 2003. p. 181–213.
30. Cleeremans A, Jimenéz A. Implicit learning and consciousness: a graded, dynamic perspective. In: French R, Cleeremans A, editors. *Implicit learning and consciousness*. New York: Psychology Press; 2002. p. 1–40.
31. Other characteristics of memory traces can also influence whether they become conscious, such as their stability in time, as well as their distinctiveness—which means whether or not they overlap with other traces.

32. Cleeremans A. The radical plasticity thesis: how the brain learns to be conscious,. *Front Psychol* 2011;**2**(86). https://doi.org/10.3389/fpsyg.2011.00086.
33. McClelland JL. On the time-relations of mental processes: an examination of systems in cascade. *Psychol Rev* 1979;**86**:87–330.
34. Dehaene S, Changeux JP, Naccache L, et al. Conscious, preconscious, and subliminal processing: a testable taxonomy. *Trends Cogn Sci* 2006;**10**:204–11.
35. Jackendoff R. *Consciousness and the computational mind*. Cambridge (MA): MIT Press; 1987.
36. Wilkenson L, Shanks DR. Intentional control and implicit sequence learning. *J Exp Psychol Learn Mem Cogn* 2004;**30**:354–69.
37. Fu Q, Fu X, Dienes Z. Implicit sequence learning and conscious awareness. *Conscious Cogn* 2008;**17**:185–202.
38. Norman E, Price MC, Duff SC, et al. Gradations of awareness in a modified sequence learning task. *Conscious Cogn* 2007;**16**:809–37.
39. Metcalfe J. Feeling of knowing in memory and problem solving. *J Exp Psychol Learn Mem Cogn* 1986;**12**:288–94.
40. Mangan B. Taking phenomenology seriously: the 'fringe' and its implications for cognitive research. *Conscious Cogn* 1993;**2**:89–108.
41. Mangan B. Sensations ghost: the non-sensory 'fringe' of consciousness. *Psyche* 2001;**7**(18). http//:psyche.cs.monash.edu.au/v7/psyche-7-18-mangan.html.
42. Isenman L. Towards an understanding of intuition and its role in scientific endeavor. *Perspect Biol Med* 1997;**40**:395–403.
43. Dienes Z, Scott R. Measuring unconscious knowledge: distinguishing structural knowledge and judgment knowledge. *Psychol Res* 2005;**6**:338–51.
44. Price MC, Norman E. Intuitive decisions on the fringe of consciousness: are they conscious and does it matter? *Judgm Decision Making* 2008;**3**:28–41.
45. Scott R, Dienes Z. The conscious, the unconscious, and familiarity. *J Exp Psychol Learn Mem Cogn* 2008;**34**:1264–88.
46. Damasio A. *Descartes' error: emotion, reason and the human brain*. New York: G. P. Putnam's & Sons; 1994.
47. Dienes Z, Perner P. The cold control theory of hypnosis. In: Jamieson G, editor. *Hypnosis and conscious states: the cognitive neuroscience perspective*. New York: Oxford University Press; 2007. p. 293–314.
48. Lau HC, Passingham RE. Unconscious activation of the cognitive control system in the human prefrontal cortex. *J Neurosci* 2007;**27**:5805–11.
49. Pothos EM. Expectations about stimulus structure in implicit learning. *Mem Cogn* 2005;**33**:171–81.
50. Ziori E, Pathos EM, Dienes Z. Role of prior knowledge in implicit and explicit learning of artificial grammar. *Conscious Cogn* 2014;**28**:1–16.
51. Dienes Z, Longuet-Higgins C. Can musical transformations be implicitly learned? *Cogn Sci* 2004;**28**:531–8.
52. Leung L, Williams J. Crosslingisitic differences in implicit language learning. *Stud Second Lang Acquis* 2014;**36**:1–23.
53. For a review of current work on synesthesia, which may be a rewarding area of study with respect to this question, see Ward J. Synesthesia. *Annu Rev Psychol* 2013;**64**:49–75.

# CHAPTER 3

# Intuitive Cognition and Intuition

## Contents



Scientific endeavor has sometimes been divided into two sequential phases called discovery and verification.[1] Linus Pauling's intuition about the biochemical mechanism behind hereditary disease illustrates intuitive cognition in its more acknowledged role, the generation of hypotheses during scientific discovery Pauling's extraordinary intuition occurred in 1945. Watson and Crick's solution to the structure of DNA was still 8 years in the future, and it was not yet known that the function of genes was to encode proteins. Crick credited Pauling's bold hypothesis as the germinal insight behind molecular biology.[2]

## INTUITIVE COGNITION IN SCIENTIFIC DISCOVERY: IT MUST BE A "MOLECULAR DISEASE"

A colleague mentioned to Pauling the recent finding that red blood cells from patients with sickle cell anemia take on an elongated or sickle shape in the low oxygen of venous blood and then recover their normal round shape in the higher oxygen of arterial blood.[3] Almost instantly Pauling responded


*Understanding Intuition*
ISBN 978-0-12-814108-3
https://doi.org/10.1016/B978-0-12-814108-3.00003-8

that sickle cell anemia must be a "molecular disease." He realized it had to be a disease of the oxygen-carrying hemoglobin molecule, which is caused by a small error in the genetic material that he surmised must encode this oxygen-carrying protein. The error, he predicted, generates a small change in the surface of hemoglobin molecules after they have discharged their oxygen. This surface change then allows the hemoglobin molecules to bind together in the low oxygen of venous blood into long rigid rods, which act from the inside to distort the usually round red blood cell into a crescent shape. Judson, who chronicled the early days of molecular biology in *The Eighth Day of Creation,* called it an astonishing instant of physical intuition even for Pauling, who was well-known for his physical intuition.[3]

Pauling's sweeping synthesis depended on the intuitive integration of a wide range of stored knowledge, along with his sense that something must reversibly deform the membrane of the diseased cells. Red blood cells contain little but water and hemoglobin. Much of Pauling's work before the war had focused on the chemical and structural properties of hemoglobin and the changes they undergo with oxygen binding. He had observed that hemoglobin molecules could bind together to form long, hard, needle-like structures. Currently, he was studying various intermolecular forces, such as van der Waals forces, hydrogen bonding, and other weak forces, which he had proposed bind macromolecules to each other.

He had reason to think that hemoglobin was made under genetic control. He knew that hemoglobin from one species causes immune reactions in other species. He had also observed that hemoglobin from different species formed different-shaped "needles." In addition, he recognized that the surface alteration of hemoglobin, and therefore the change in the genetic material, was likely small because the sickled cells revert to their normal shape at high oxygen tension.

Pauling felt certain of his sweeping synthesis because it fit with the whole mesh of knowledge he had about the hemoglobin molecule. In the next few years, he and a coworker used electrophoresis to demonstrate that there was small difference in charge between normal and sickle cell hemoglobin.[4] They also showed that all of the hemoglobin was altered in those with the full-blown disease and half of it in asymptomatic carrier parents, who would have only one copy of the defective gene. This confirmed that the disease had a molecular basis and also that it was encoded in the genetic material.

The findings were also pivotal in another way. After the actual structure of DNA was solved a number of years later, Ingram, a worker in Francis

Crick's lab, showed that sickle cell hemoglobin indeed has only a single amino acid change from the normal molecule: a hydrophilic amino acid replaces a hydrophobic one.[5] This study provided the first confirmation that a specific genetic mutation could lead to a change in the amino acid sequence of a protein, as predicted by Pauling, and as permitted by the DNA model. Pauling, it turns out, did get one detail wrong. Sickle cell hemoglobin does not form the ordered needles in the absence of oxygen he first envisioned. Instead, it forms much more disordered, or denatured, strands that nonetheless distort the red blood cell from within, just as he predicted.

As Pauling's insight demonstrates, intuition in the context of scientific discovery depends largely on intuitive cognition, or putting together a variety of different facts in an interwoven, or parallel-interactive, way. This integration most often incorporates novel and/or untested assumptions. However, on occasion it may instead combine novel collections of *known facts* in a parallel-interactive fashion. In all cases, it leads to novel predictions that require verification.

## INTUITIVE COGNITION IN SCIENTIFIC VERIFICATION: THE FORM AS WELL AS THE CONTENT OF THE DATA WAS STRIKING

Intuitive cognition sometimes also participates in an important way in verifying or justifying scientific hypotheses. I will illustrate this less recognized role for intuition-related phenomenon with an experience I had in the course of my own work as a biologist. It also helps to illuminate the often indistinct boundary between higher-level intuitive cognition and intuition proper as it relates to scientific analysis.

As a graduate student, I studied the movement of digestive enzymes from the pancreatic cells where they are made to the small intestine where they digest the food we eat. Observations from the laboratory in which I worked suggested that digestive enzymes can exist in the watery interior of the pancreatic cells and can pass directly through a variety of its membranes.[6] This ran counter to the well-accepted view that all digestive enzymes are packaged into storage granules as they are synthesized, that they cannot exist in soluble form in the cell, and that they do not cross membranes. Instead of the presence of only one fixed and one-way pathway for digestive granules, as everyone thought, our findings were consistent with a much more complex model featuring a number of parallel paths interconnected by bidirectional fluxes of soluble digestive enzymes across the various membranes of the cell. I suspect my intimacy with this complex array of interconnected pathways enhanced my receptivity to the data from the experiment I will describe.[7]

I continued to explore the movement of proteins and parts of proteins (peptides), through membranes as a postdoctoral fellow. In a different cell type, my collaborator, J. Fred Dice (called Paulo), and I showed that some breakdown fragments of proteins were not necessarily degraded further into single amino acids, as almost everyone assumed. Instead they could cross the outer membrane of the cell into the surrounding medium.[8] The likely intracellular source of the fragments was an organelle named the *lysosome,* whose function is to degrade many of the cellular proteins.

We needed to confirm that the protein fragments we saw exiting the whole cells were in fact produced in lysosomes, crossed its membrane into the intracellular fluid, and then crossed the cell membrane itself (Fig. 3.1). In what we thought was only a preliminary experiment, we loaded the cells with radioactively tagged proteins, isolated large amounts of

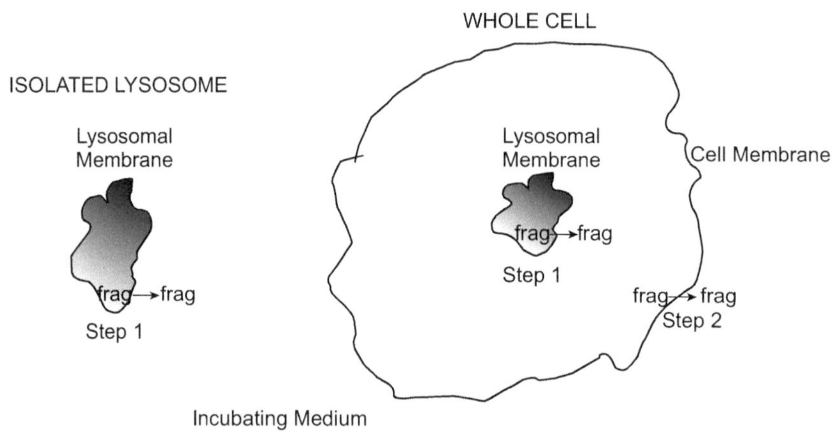


**Figure 3.1** ***Release of specific peptides from isolated lysosomes.*** In an earlier experiment, we had observed that specific breakdown fragments (frag) of radiolabeled proteins exited from whole cells. To show that the process happened in two steps (release from the lysosomes into the watery interior of the cell and then release from the cells), we added the radiolabel proteins to the cells, isolated the lysosomes, and incubated them at two different temperatures for 2 h. We then determined which fragments were released into the incubating medium and which were retained by the lysosomes.

lysosomes, and incubated one portion at one temperature and another at a different temperature. When I characterized the protein fragments that were released into the incubating medium and those that were retained by the organelle at each temperature, I was astounded. The two sets of graphs that visualized the data offered conclusive evidence that the outer membrane of the lysosome was completely permeable to the specific small protein fragment of most interest to us (Appendix 1, Fig A1.1). They also revealed a much more subtle, complex, and convincing story about the membrane's permeability to only certain fragments than we had any reason to suspect.

Strikingly, I did not have to figure any of this out; it was simply the story the four highly entwined graphs told me almost immediately. The directness of this knowledge made it one of the most grounding experiences I have had as a scientist. I felt as if I were inside the cell seeing the intracellular dance. The data was also the most convincing I have encountered. Its form, as much as its scientific import, generated a profound feeling of respect in me. On the one hand, it was complex because it detailed a number of different processes that had to be considered in the context of each other. On the other hand, because these processes all resolved themselves into such an internally consistent and self-reinforcing whole, it was very simple.

After repeating the experiment and confirming the results, I shared the data from both experiments with Paulo. He seemed as pleased as I was with the results. However when I presented my findings at a laboratory meeting a number of days later, my other coworkers appeared to be deeply puzzled. Eventually one commented that the peak that I claimed corresponded to the fragment in which we were primarily interested, the most abundant one, was somewhat broadened. Therefore it might not represent the same one we had previously shown to exit the intact cells in large amounts. When Paulo said, "They seem skeptical," I did not say anything to defend my understanding of the data.

I knew that the four graphs taken together provided a number of additional levels of support for my claim, but I did not think to point out this evidence. It seemed to me that the only way for my coworkers to comprehend the results was to look at them more. Eventually I recognized that Paulo shared their skepticism. When I reminded him of his prior enthusiasm for the data, he seemed unable to recall this earlier response. I knew my colleagues were just doing their job by scrutinizing the results so carefully, especially given their knowledge of my long-standing bias in favor of the permeability of membranes to proteins and protein fragments. Yet part of what impressed me so much about the results when I first saw them was that they had an inherent integrity transcending any bias I might have.

Several hours later, Paulo asked to see the data again. After a moment or two, he said, "What great data!"[9] I was, of course, relieved that the inexplicable cognitive chasm between us had been breached, but in another sense I was even more dumbfounded than before.

After we repeated the experiment a number of times, we wrote the work up and sent it off for publication.[10] We characterized the raw data and stated the most dynamic level to the conclusion, but we did not attempt to explain the rest of it in detail. Instead we focused only on how the selective permeability of the membrane to different fragments could account for the results. Nonetheless I was aware that, even if I had wanted to, I would not be able to make a clear statement about the various levels to the conclusion, nor point to the evidence backing up each level. Each aspect of the data was too enmeshed with every other to address it in a systematic way.

## Making Connections

I began to make the connection between my strange experience presenting the lysosomal data to my colleagues and intuitive cognition the following year as I was preparing a talk on these experiments. I felt I should present the results in a more linear, or stepwise fashion, because my audience was primarily nonscientific—my colleagues at the Bunting Institute, an interdisciplinary institute. I forced myself to begin to untangle all of the different layers to the conclusion and all the different strands of evidence supporting each layer.

A short way into this arduous task, it suddenly struck me that intuitive cognition, or intuitive thinking, depends on exactly this complex intertwining of strands of evidence, layers of conclusion, and internal tests for consistency. It differs from ordinary reasoning because it entails processing a variety of different but interrelated strands of information simultaneously, or in *the context of each other*. Moreover it often results in a variety of entwined conclusions and internal controls, which means that the same piece of information can have multiple roles. This densely interwoven quality is what gives intuitively processed information the sense of certainty it conveys.

I also recognized that because of this dense interweaving, my conscious perception of the complete raw data, its integration, and its evaluation had occurred more or less simultaneously. The different levels of relationship and interrelationship had all condensed into a coherent web of understanding. The most dynamic relationship stood for the whole web in my mind, much the way a Japanese haiku captures a complex dynamic in 17 syllables. This kind of interwoven integration of information, which most often occurs in the unconscious, is at the core of intuition.

The convergence of multiple interactive characteristics into a *familiar* pattern accounts for much of the way the unconscious mind puts the world together. It allows us to "just know" based on past experience that *this* is a *this*, or *that* means *that*. The patterns are flexible and consist of multiple soft, or malleable, strands of information, which allows *knowing* despite inherent variation and partially obscuring conditions. Integration may sometimes require bidirectional processing, or a back-and-forth between several alternatives, either above or below awareness, to find the best fit.

The established patterns that account for much of lower-level cognition, such as recognizing objects, tend to generalize readily. We recognize all breeds of dogs as dogs, with little problem. The patterns responsible for *some higher-level* cognition, such as many aspects of expertise, can also be applied to many situations. Yet the completely novel patterns responsible for others examples of higher-level cognition are tightly linked to the details of a specific situation, such as the pattern formed by the lysosomal data. They are thus much less likely to generalize to other situations. Some people seem to depend on parallel-interactive processing for higher-level cognition more than do others.

The question of whether the use of parallel-interactive processing for novel *higher-level* cognition is the same as intuition is a challenging one. On the one hand, if the processing happens below awareness and leads to novel and significant conclusions, it would seem to qualify as intuition. On the other hand, some intuitive processing clearly occurs at least partly at a conscious level. My experience when I first saw the data I described earlier felt like rigorous conscious integration of the data instead of intuition. Not all of the processing occurred at a conscious level. But enough did so I was certain my conclusions were rigorous and that others would also be able to understand them.

The distinction between intuitive processing and *intuition proper* may remain fuzzy even if the integration occurs completely consciously. The cognitive chasm that opened between my colleagues and myself when I presented the lysosomal data illustrates the problem. What seems like rigorous intuitive integration of the evidence to one person may appear unintelligible to others, or to make assumptions that require verification.

## Barbara McClintock and Jumping Genes

The ambiguity between intuitive processing and intuition is highlighted by the work of Barbara McClintock.[11] Like Pauling, she was a Nobel laureate whose cognitive style depended heavily on intuitive integration. Also like Pauling, her work made groundbreaking contributions to understanding the genome.

McClintock trained as a geneticist and cytologist in the 1920s. As a graduate student, she developed a histochemical technique to identify the nine different chromosomes of maize, or corn, under the microscope. For most of her career, she planted, pollinated, and harvested generation after generation of maize, correlating alterations in the color and markings of individual kernels with chromosomal changes she observed under the microscope. As a young professor, along with a graduate student, she documented crossing-over of chromosomal fragments during meiosis and recombination of the genetic material.

Most geneticists at the time believed that the exclusive function of all genes was to code for physical traits. They also believed that the position of specific genes on the chromosomes did not change from generation to generation. However McClintock observed complex patterns of change during successive generations in the coloration of different kernels from ears of corn produced from a single kernel. From these patterns of change, she deduced that the genome included genes that control the expression of other genes. This indicated a new function for genes—to control other genes. Moreover her studies revealed that some of these control elements themselves change position on the chromosome.

McClintock's work is extremely difficult to follow, partly because of her dependence on intuitive cognition. Each new observation rests on the integration of multiple past layers of understanding. Most outside her immediate circle did not understand or appreciate her findings. Some even questioned her sanity as her observations became increasingly out of the mainstream. For a while she stopped publishing her findings because of the resistance they produced.[12] Yet as one of her students noted, McClintock could *see* the chromosomes switching on and off as she looked at her slides under the microscope.[11] She could also teach those who were open to her ideas to *see* as she did.

McClintock's studies on the existence of controlling elements received the recognition they deserved only after a similar phenomenon was described in bacteria in the 1960s. Her findings on mobile genetic elements, or transpositions (now called *transposons* or *jumping genes*) were not appreciated until later. In 1983 she was awarded the Nobel Prize for her discovery of transposition.

Transposons have been found in every species. They are probably more likely to arise during periods of stress,[13] and they can function in an adaptive manner.[14] Growing evidence also points to an important role for them in evolution.[15]

Speaking about her ability to grasp the pattern of genetic expression on each ear of corn she grew, McClintock told her biographer that she immediately understood every plant without being aware of the integration. When asked what *understanding* means in this context, McClintock captured the completely rigorous feeling intuitive cognition often conveys. Her mind, she answered, was like a computer that functioned in an almost instantaneous and completely trustworthy way.[11(p102)]

Yet this rigor is not the kind associated with ordinary computers. Talking about her cognitive process with regard to an earlier and less controversial discovery, McClintock pointed to the unusual logic of the whole that structures the deeper sort of understanding, or mental seeing, often associated with intuitive cognition and intuition and, at the same time, accounts for the certainty it brings:

> *Why was I so sure they were ring chromosomes? I could convince anyone to call them ring chromosomes before anyone had seen them. That was, God knows, true confidence. I was not trying to persuade anybody, but I was convinced. Why was I so convinced that the thing* had to be, *that it* couldn't *be anything else but that?... The logic was compelling. The logic made itself, the logic was it. What's compelling in these cases is that the problem is sharp and clear. The problem is not something that is ordinary, but it fits into the whole picture, and you begin to look at it as a whole.... Its what goes on in the whole cycle. So you get a feeling for the whole situation of which this is [only] a part.*[11(p67)]

## INTUITION AND BRAIN LATERALITY

The right and left hemispheres of the brain process information differently, which suggests one possible approach to understanding the tendency of some people to favor intuitive cognition for higher-level thought. The right brain–left brain distinction has become part of popular culture, and it may be the only thing some people know about brain function. Despite its buzzword status, it continues to be an active area of investigation.

The right brain is generally associated with a global perspective, a sense of the whole, or the big picture. The left brain, in contrast, tends to be associated with a local focus, interest in the details, and logical analysis. Most assume that intuition depends more heavily on the right brain than the left, and it is frequently linked with the idea of a holistic or global perspective.

Early research examining the difference between the two hemispheres located different mental abilities exclusively to one hemisphere or the other. The left hemisphere was seen as the site of language processing, whereas visuospatial processing seemed to occur exclusively in the right hemisphere. Subsequent research, however, has revealed a more complex picture. Some aspects of speech depend on the right hemisphere, and both sides of the brain cooperate in certain aspects of visuospatial processing. For example, the right hemisphere is critical to understanding prosody—the rhythm, intonation, and emphasis of speech.[16] Both microscopic and macroscopic anatomical differences have been observed in the structure of the two hemispheres, which can help account for their functional differences.[17]

Brain lesions caused by tumors or strokes in specific higher-level regions of the right or the left hemisphere, by degrading or eliminating the perspective of the injured hemisphere, can highlight the underlying perspective of the intact hemisphere. The line drawing shown on the left in Fig. 3.2 is often used to test patients with brain lesions. The center shows an attempt by a patient with a left hemisphere lesion to reproduce the drawing, and the right, by a patient with a right hemisphere lesion. The patient with the largely unopposed right hemisphere, owing to left hemisphere damage (LHD), focuses on the outlines of the figure and for the most part ignores the details. In contrast, the patient with the intact left hemisphere lesion, owing to right hemisphere damage, focuses on the most salient details and gives little attention to the allover form.

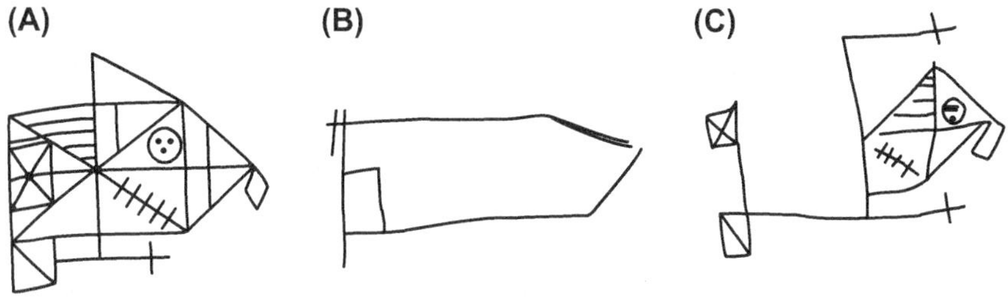


**Figure 3.2** ***Patients with right or left hemisphere damage reproduce a complex line drawing from memory.*** (A) A line drawing used to test patients with brain lesions. (B) The patient with a left hemisphere lesion and intact right hemisphere draws only the global form when trying to reproduce the figure from memory. (C) The patient with a right hemisphere lesion and intact left hemisphere draws many local elements, yet they are in incorrect relationship to each other. *(Reproduced from Robertson, Lamb.* Cogn Psychol *1991;**23**:299–330. Reprinted with permission from Elsevier.)*

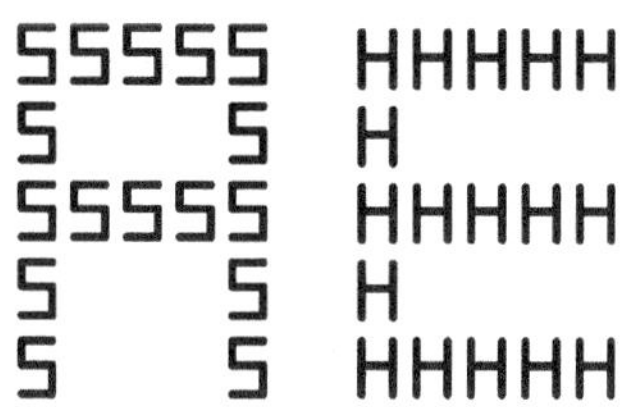


**Figure 3.3** ***Recognition of different levels of structure by patients with right and left hemisphere damage.*** The large A is composed of a number of small S's and the large E is composed of many small H's. A patient with a right hemisphere lesion takes less time to recognize the S's than the A, for example. The reverse is true in patients with damage to the left hemisphere. *(Adapted from Lamb, Robertson, Knight.* J Exp Psychol Learn Mem Cogn *1990;**3**:471–83.)*

A similar phenomenon is shown in Fig. 3.3. The large A, for example, is composed of a number of small S's. A patient with a right hemisphere lesion takes more time to recognize the A than an S, whereas in patients with damage to the left hemisphere, the reverse is true. A similar effect can be shown with completely normal subjects when the figure is shown to their extreme left or right, because the right-most portion of the visual field of both eyes feeds into the left cerebral hemisphere, and vice versa.

Another striking example of the differences between the right and left hemispheric processing of the same stimuli is shown in Fig. 3.2C. Equilateral triangles point equally in three directions. When subjects with either right or with left hemisphere damage are shown a single equilateral triangle, they mistakenly report it as pointing in one direction or the other with about the same frequency as normal subjects do (graph on bottom left of Fig. 3.4). However, when the triangle is aligned with two others, as shown in Fig. 3.4 on the right, both groups of patients tend to respond differently than normal subjects (graph on the bottom right of Fig. 3.4). Those with LHD are more likely than normal controls (CTL) to see the middle triangle as pointing in the direction in which the three triangles are aligned. The unopposed right hemisphere with its focus on the object in its surrounding context is more likely to sway the brain than in normal subjects. The opposite is true with patients with right hemispheric damage. With only left hemispheric

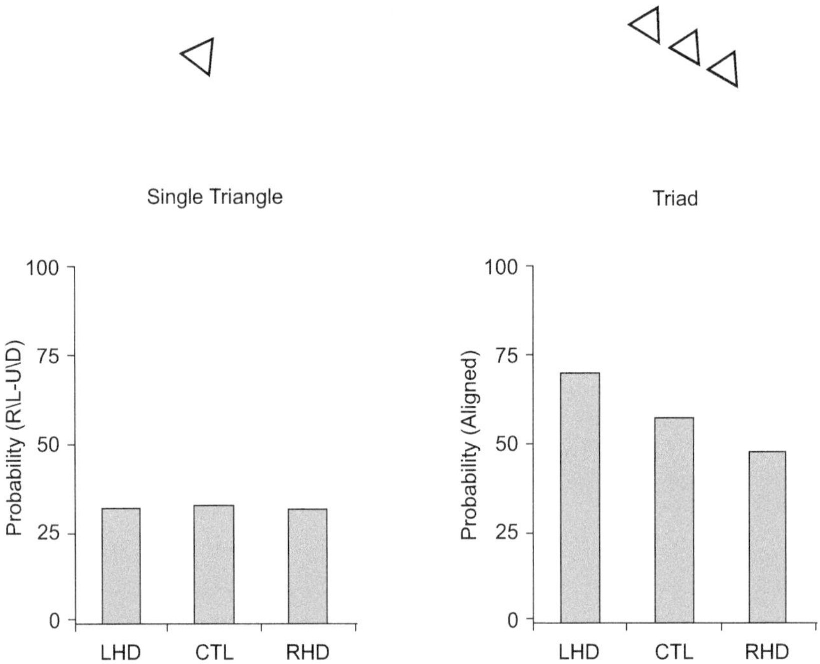


**Figure 3.4** ***Patients with left hemisphere damage (LHD) and right hemisphere damage (RHD) respond the same as controls to a single equilateral triangle but differently when it is aligned with others.*** (*Left*) When subjects with either right hemisphere damage (RHD) or left hemisphere damage (LHD) are shown a single equilateral triangle, they make the mistake of reporting it as pointing in one direction or the other with about the same frequency as do normal control subjects (CTL). (*Right*) However, when the triangle is instead aligned with two others, both groups of patients respond differently from normal subjects. Those with LHD are more likely than normal controls to see the middle triangle as pointing in the direction in which the three triangles are aligned. The opposite is true with patients with RHD. *(Adapted from Roberson LC, Ives R.* The two sides of perception. *Cambridge (MA): MIT Press; 1998. Reprinted with permission of MIT Press.)*

function, they are more likely than are normal controls to overlook the surrounding context.

In most people, the mind integrates both the global and local perspectives. The capacity of the brain to integrate two different views of the environment allows for a considerably richer and more flexible understanding than any single view alone. However this integration is not necessarily evenly balanced. Some may tend to pay more attention to local information and others to global. Genetic factors have a role and cultural factors are also important.[18,19]

The division of labor between the two hemispheres in normal subjects is also more complex than these experiments suggest. What corresponds to a local focus in one moment might become a global focus in the next. To explain the ability of both hemispheres to work over the same range of scales with both hearing and vision, Ivry and Robertson proposed a two-stage process.[20] The brain first decides the scale of spatial frequencies, or densities, that is appropriate to the demands of the task. The right hemisphere then filters out the high-frequency, or high-density, information and amplifies the low-frequency, or low-density, information, which can reveal larger patterns. The left hemisphere, in contrast, filters out the low-frequency information and amplifies the high-frequency information, which can help reveal the details.[21]

An interesting study used functional magnetic resonance imaging to localize the differential activity associated with narrowing the spotlight of visual attention, or "zooming in," to a specific region of the left hemisphere (the left anterior intraparietal sulcus).[22] In contrast, it localized the differential activity associated with expanding the spotlight of attention, or "zooming out," to a specific region of the right hemisphere (the inferior frontal gyrus). Interestingly, the same region of the right hemisphere that is activated by "zooming out" is also activated when subjects are asked to divide their attention and follow changes in three aspects of a given stimulus simultaneously.

These studies are consistent with the idea that the right hemisphere focuses more on the global level and the left hemisphere on the details. However the left hemisphere is not always more active or faster in recognizing individual details.[23] One possible explanation for the ability of the right hemisphere to attend better to the details in some situations is that the two hemispheres encode different kinds of relationships. Kosslyn and his colleagues argued that in the spatial realm the right hemisphere encodes *coordinate,* or metric, relationships, which give precise point-to-point information.[24] In contrast, the left encodes categorical, or more general or abstract, relationships that hold across all examples of a class. This includes relationships that are invariant under spatial transformations—for example, what allows us to recognize a hand as a hand when it is presented from many different angles. By the same token, the left hemisphere also encodes more general categorical spatial relationships, such as *on* and *off*, or *in front of* or *behind*. The researchers also suggested that the left hemisphere is more likely to encode basic-level category members, such as a chair, whereas the right hemisphere is more likely to encode specific exemplars or members of a category, such as a *rocking chair* or a *bean bag chair*.[25]

## LINEAR VERSUS RELATIONAL LOGIC

Words such as *global*, *holistic*, and the *big picture* that are associated with right hemisphere function are sometimes used as substitutes for intuition in popular culture. They are certainly correct to use in the context of intuition. However, unexamined they may obscure as much as they clarify this elusive mental ability. They veil some illuminating complexity about the underlying mechanism of higher-level intuitive thought. Another look at the situation I described earlier with my colleagues can provide a richer, more comprehensive understanding of intuitive cognition and intuition.

One possibility is that given my left-handedness and intuitive bent, and thus possible right hemispheric bias,[26] I received a coarse-grained sense of the data that did not include all of the salient details. The idea that normal people with a right-hemisphere bias focus primarily on the outlines of a situation, whereas those with a left-hemisphere bias focus primarily on the details, come from extrapolation of data from brain-damaged patients, such as that shown in Fig. 3.2, to the cognitive activity of normal subjects. However such an extrapolation is not valid—especially in the context of a scientific investigator trying to interpret their data. The explanation for the cognitive chasm I experienced with my colleagues likely lies in a more nuanced understanding of the effect of brain laterality on higher-level cognition, one that resonates with several of the perspectives mentioned in the previous section.

I will begin with my colleagues. What must have happened, I realized long afterward, was that they "zoomed in," or narrowed their focus of attention, first to the graph, and then to the peak that seemed most important, the one I said represented the fragment of interest (Fig. A1.1, peptides release into incubating medium at 24 degrees). They then applied ordinary logic to assess the validity of my conclusions. Because they saw my attribution of this critical peak as questionable, they saw any conclusions I drew from the data to be questionable as well. With ordinary logic, which depends on rule-based and sequential processing, if the central assumption is uncertain, no meaningful inferences can be drawn until it is clarified. This traditional sequential, analytic approach works well much of the time.

I also started by focusing on this graph (which I drew first) and its seemingly critical peak. However, when this proved inconclusive, I broadened my attention, or "zoomed out," to consider all of the graphs simultaneously (Appendix 1, Fig. A1.1). I then used an alternative form of logic, a parallel-interactive or contextual one, to make sense of the data. The ambiguity my coworkers stressed did not weaken the overall import of the data when all of the peaks and troughs in the four graphs are considered in the context of each other. Taken together, they demonstrate a number of interrelationships, which, in addition to strengthening the attribution of the peak in question, point to a multilevel conclusion. At the same time, they offer a number of internal controls that vouch for the validity of the different levels of this multilevel conclusion. In Appendix 1, I have attempted to reproduce this reasoning. Appendix 2 attempts a sequential analysis of the data. It shows that the results taken together can be understood with ordinary logic, if at all, only with great difficulty.

My problem was that when my colleagues challenged me, I could not undo the intuitive integration that had occurred in my mind when I first saw all of the data together. Once I saw how well all the peaks and valleys in the four graphs fit together, I could neither undo the integration that had occurred nor comprehend that others might approach it *only* in the traditional manner. Moreover I could not understand that my collaborator Paulo could approach it in either way, but apparently only one way at a time.

At least with respect to these data, my colleagues and I were living in different cognitive universes that depended on different "rules." Theirs depended on the straightforward analytic and reductive logic that governs much of scientific discourse. Understanding depends on separating out the critical component or components, considering them in great detail largely in isolation from the rest of the system, and determining what is (or is not) true about them.

In contrast, my universe at that moment depended on parallel interactive or contextual logic. The soundness of any piece of information does not rest on itself alone. Rather it is shared or distributed among all the other information. Validity depends on mutually reinforcing interactions with the other components. Looked at in this way, the issue is less a question of a local or a global conceptual focus than it is the way the mind considers the details. One cognitive mode seeks to understand the details in a rule-based or categorical and *context-independent way*, and the other seeks to understand them in a simultaneous, parallel-interactive, and *context-dependent* way.

The ability to break down situations into their components and consider the most important element or elements in isolation is a specialty of the conscious mind. This is consistent with the idea that consciousness is limited by working memory, which appears to hold only a limited number of items ($7 \pm 2$) at the same time.[27] The unconscious mind does not have this limitation. It can integrate many items in parallel, in part by keeping track of their mutual interactions.

In retrospect, presenting these results to my coworkers was an invaluable experience. It greatly magnified what is probably only a subtle difference in the way those who favor intuitive cognition and those who do not sometimes put their worlds together. Although the integration I performed when I first saw the two sets of graphs occurred partly at a conscious level, the extended experience helped externalize as well as concretize the kind of intuitive integration that often occurs completely below awareness.

## COGNITIVE ASSUMPTIONS UNDERLYING THE TWO ASPECTS OF MIND

The conscious mind tends to focus attention on the most important or pronounced elements. In contrast, the unconscious mind tends to favor intuitive or parallel-interactive logic, which depends on the pattern of connections that develop between all of the elements. Consequently, it is tempting to claim that the left hemisphere is associated with conscious mental activity, with its limited processing capacity, whereas the right is associated with unconscious mental activity, with its much larger processing capacity. However consciousness is a product of both hemispheres.

Adding a distinction between conscious *thought* on the one hand and conscious *experience* on the other allows the following generalization. The left hemisphere with its propensity for serial logic and thinking about one thing at a time may be relatively *more* associated with conscious thought. Conversely, the right hemisphere with its propensity to process many items in parallel may be relatively *more* associated with unconscious thought.

Evolution appears to have provided different cognitive assumptions or filters to these two components of mind. They orient the mental faculty and underlie its method of approach. However they reverse the linkage between the right hemisphere and the global and the left hemisphere and the details, suggested by the brain laterality studies of brain-damaged patients!

The cognitive assumption underlying *unconscious* information processing, to the extent it is dominated by the right hemisphere, seems to be that only the parts, or the details, and their potential connections exist. Unconscious information processing tends to move out from the specific details and their potential interconnections to the whole. What it knows of the whole emerges from the parallel-interactive integration of the parts into something meaningful, guided by the emotional/motivational system. This is why the kind of artificial information processing that corresponds to unconscious processing is called *bottom-up,* or sometimes *emergent, processing.* Basic visual understanding, for example, depends on the emergent connectivity of lines, brightness, and color to create a meaningful visual scene.

This apparent reversal of the assumptions underlying unconscious information processing—at least when guided by the right hemisphere—is the key to a paradox at the heart of intuitive cognition. Intuitive cognition starts with the parts, or the details, and their interactions and focuses through to the whole. However intuitive, or parallel-interactive processing of the parts and their interactions happens largely in the unconscious. Therefore the right hemisphere is more likely to be associated with a sense of the whole or the big picture, since it is often the only thing that is perceived at a conscious level. If this whole provides *significant and novel* understanding, it corresponds to an intuition.

In contrast, the underlying cognitive assumption behind conscious information processing, at least when controlled by the left hemisphere, is that an a priori big picture or global whole already exists. The left hemisphere then seeks to understand or confirm the big picture from an examination of the parts or the details. It moves in from the whole to its parts.

In contrast to the context-dependent relational approach, in which the logic is inherent in the subject matter, the logic is imposed from the outside in what is sometimes called a "God's eye perspective."[28] Artificial information processing that mimics this aspect of conscious functioning is called *top-down information processing.*

The importance of the implicit idea of a big picture is easy to see when the logic imposed is deductive and moves from the general to the particular. However, the sense of a big picture is often no more than an a priori unconscious assumption that a context in fact exists for the details. For example, the process of induction requires an implicit assumption that the items that one groups together fit into an ordered and meaningful larger context, as well as some tentative assumptions as to what this context might be.

Split-brain studies by Michael Gazzaniga and coworkers support the idea that the cognitive assumption behind the functioning of the left hemisphere is an implicit idea of the whole.[29] To control the spread of seizures in people with severe epilepsy, neurosurgeons sometimes sever the corpus callosum, the major connection between the two hemispheres, creating a "split brain." In split-brain subjects, the two sides of the brain function largely independently. Different visual information can be presented to the left and the right sides of the brain simultaneously by having subjects fix their eyes on a central point. As Fig. 3.5 illustrates, researchers showed the left hemisphere of a split-brain subject a picture of a chicken claw and at the same time showed the right hemisphere a snow scene. The right hemisphere controls the left side of the body, and the right hemisphere, the left side of the body. When asked to choose a related card with each hand from a panel of cards in front of him, some only his right hemisphere could see and some only his left, the subject's right hand (controlled by the left hemisphere) correctly picked the chicken as the match for the chicken claw, and the left hand (controlled by the right hemisphere) correctly picked the snow shovel as the match for the snow scene.

However when he was asked why he picked the shovel with the left hand, he reported that the shovel was needed to clean out the chicken coop. Speech is generally under the control of the left brain, which was privy to only part of the information. Although the speaking part of his mind and his left brain had no knowledge of the snow scene or the snow shovel, it nonetheless came up with a story that incorporated the shovel in an apparently logical fashion. The left hemisphere assumes that an external context and meaning exists for whatever details with which it is

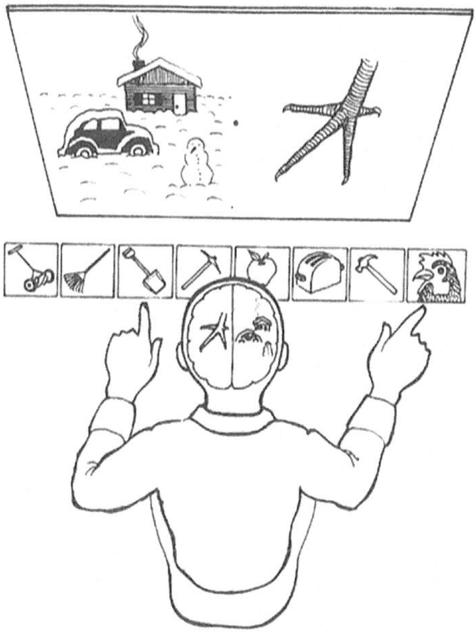

**Figure 3.5** ***The left hemisphere's implicit sense of a whole.*** Researchers showed the left hemisphere of a split-brain subject a picture of a chicken claw. At the same time, they showed the right hemisphere a snow scene. He was asked to pick a relevant card with each hand from a panel of cards in front of him, some of which only his right hemisphere could see, and some only his left. The right hand (controlled by the left hemisphere) correctly picked the chicken as the match for the chicken claw, and the left hand (controlled by the right hemisphere) rightly picked the snow shovel as the match for the snow scene. However, when he was asked why he picked the shovel with the left hand, he reported that the shovel was needed to clean out the chicken coop. Speech is generally under the control of the left brain, which was privy to only part of the information. *(Reproduced from Gazzaniga MS, Le Doux JE.* The integrated mind. *New York: Spring; 1978. Reprinted with permission of Springer.)*

presented. Left to its own devices, it tends to make up one that seems to fit. Gazzaniga has labeled it "the interpreter"[30] because of its tendency to extract patterns from what it encounters. Also, it can reason on its own (albeit with greater or lesser success) whereas the right hemisphere cannot.

Speech itself is a good example of a top-down activity. We speak a single word at a time. However, to make sense grammatically, each word in a sentence or unit of speech has to presuppose knowledge by the speaker of the whole unit of speech. This hidden a priori assumption of a whole supports the serial and grammatical output required by speech.

## PATTERN RECOGNITION, BRAIN LATERALITY, AND INTUITION

The apparently topsy-turvy view of the underlying importance of parts and wholes in right- and left-hemisphere function I have suggested can sort out a common confusion about pattern recognition, brain lateralization, and intuition. Because pattern recognition depends on parallel interactive connection and often has a role in intuition, it is sometimes associated more with the right hemisphere than the left. Yet others such as Gazzaniga associate pattern recognition with the left-hemisphere "interpreter."

It is useful to differentiate between two senses of pattern recognition. One consists of the recognition of often difficult to decipher novel patterns that require complex interconnection of the parts. The other consists of more familiar patterns that the mind can increasingly recognize as a whole, can extrapolate to other situations or use in a top-down, rule-based way, and can readily verbalize (or otherwise symbolize). In other words, pattern *creation/formation* is probably *more* dependent on the right hemisphere, with its ability to maintain various diverse elements in an active state and integrate them based on multiple interconnections. In contrast, recognition of repetitive patterns, as Goldberg has suggested, may become progressively more dependent on the left hemisphere over time.[31] The migration of codified patterns to the left hemisphere over time would in turn highlight the distinction between the kind of pattern recognition involved in heuristics, habitual cognition, and even much of expertise, on the one hand, and the kind involved in higher-level intuition, on the other hand, which can significantly reorder old ways of thinking.

Goldberg equates the growing ability of the left hemisphere with experience to assess situations and recognize patterns at first glance with wisdom. Certainly there is value in having a repertoire of ready solutions at hand. However, as the split-brain studies suggest, excessive dependence on left hemisphere function with its a priori assumption of a whole has its perils. Goël says of the left hemisphere interpreter, "It simply abhors uncertainty and will complete any pattern, often prematurely, to the detriment of the organism."[32] These patterns, he points out, can be of any kind, not only logical or statistical.

In contrast, Goël finds right-lateralized activity in various regions of the prefrontal cortex when the more obvious solutions do not work. These include indeterminate problems,[33,34] poorly structured problems,[35] and problems requiring set-shifting or the restructuring of perception.[36] These situations require the more fluid, often previously unpatterned or emergent connection making at which the right hemisphere excels. The potential for facile pattern recognition by the left hemisphere emphasizes that saying the first thing that comes to mind, which may often be wrong, is not necessarily the same thing as intuition.

An experiment by Beeman and Bowden underscores the right hemisphere advantage in the kind of cognition often central to insight and intuition.[37] They used a variant of the Remote Associates Test, which is often used in experiments on intuition and insight.[38] Subjects were given three words and asked to come up with another word, which when added would make commonly used compound words. For example, when given *duck*, *fold*, and *dollar*, the answer is *bill*. In this experiment, subjects were presented with a large number of successive triads. If they could not come up with the solution in a set time to a triad, the triad was removed from the computer screen, and either the solution or an another unrelated word was presented separately to the right or to the left hemisphere. Beeman and Bowden recorded how fast the solution words were read—and thus were recognized—compared to the unrelated words. Called solution-dependent priming, it provides a measure of how wide-ranging the activation of triad-associated words was in each hemisphere over time.

The researchers found that both hemispheres showed similar solution-related priming when tested 1 second after the triads were removed, which suggests that they both have access to the same store of words. But by 2 s, the left hemisphere began to demonstrate less solution-related priming, and at 7 and 15 s, progressively less. This suggests that the left hemisphere was increasingly limiting its focus to the closest, most dominant associates. The right hemisphere, in contrast, showed relatively constant solution-related priming over the 15 s, suggesting that it was maintaining weak and diffuse activation for a number of more distantly related words. Especially with more difficult triads, the right-hemisphere's diffuse encoding that can capture the overlap between remote associations is more successful.

Electroencephalogram studies using remote association triads that could be solved by either insight or conscious search also showed that the brain activity that distinguishes insight with this kind of task reflects primarily right-lateralized activity.[39] This was evident as soon as subjects began working on the task, and also in a short burst about 3 s before they came to the solution. The right hemisphere's ability to remain uncommitted in cognitively challenging situations is likely to be especially important in breakthrough intuitions, which often require holding complex issues open until an answer can emerge that takes into account all relevant information and its interconnections. The next chapter will also consider the difference between intuition and more ordinary cognition, but from a different perspective.

## REFERENCES


1. Reichenbach H. *Experience and prediction: an analysis of the foundations and the structure of knowledge*. Chicago: University of Chicago Press; 1938.
2. Crick F. *The impact of Linus Pauling on molecular biology*. Lecture at The Pauling Symposium Oregon State University; February 28, 1995. http://oregonstate.edu/dept/Special_Collections/subpages/ahp/1995symposium/crick.html.
3. This incident is taken from Judson HF. The eighth day of creation: makers of the revolution in biology. **1979**. New York: Simon and Schuster; 1979. p. 298–308.
4. Pauling L, Itano HA, Singer SJ, et al. Sickle cell anemia, a molecular disease. *Science* 1949;**110**:543–8.
5. Ingram VM. A specific chemical difference between globins of normal and sickle-cell anemia hemoglobins. *Nature* 1956;**178**:792–4.
6. Rothman SS. Protein transport by the pancreas. *Science* 1975;**190**:747–53.
7. For a discussion of the relationship of "equilibrium model" for pancreatic digestive enzyme secretion to intuition see Isenman L. Trusting your gut among other things: digestive enzyme secretion, intuition, and the history of science. *Found Sci* 2009;**14**:315–45.
8. Isenman LD, Dice JF. Secretion of intact proteins and peptide fragments by lysosomal pathways of protein degradation. *J Biol Chem* 1989;**264**:21591–6.
9. After reading an early draft of this chapter, Paulo, who has recently passed away, suggested that I mention the heated exchange we had in the hall after the seminar and before this reconciliation. I accused him of blowing in the wind and he accused me of not backing up my conclusions. Neither was able to understand what the other was saying.
10. Isenman LD, Dice JF. Selective release of peptides by lysosomes. *J Biol Chem* 1993;**26**:23856–9. The paper was accepted for publication based on the separation of the fragments by size. However the acceptance came with the condition that we show with an additional method of our choosing that the fragment released from both the intact cells and the isolated lysosomes were the same. The additional data provided another level of proof about the identity of the peptide capable of convincing everyone.
11. The discussion of McClintock and her work is taken from Keller EF. *A feeling for the organism*. New York: W. H. Freeman; 1983.
12. Letter from Barbara McClintock to J. R. S. Fincham. *The Barbara McClintock files, profiles in science*. National Library of Medicine; May 16, 1973. http://profiles.nlm.nih.gov/LL/B/B/G/C/_/llbbgc.pdf.
13. Capy P, Gasperi G, Biemont C, et al. Stress and transposable elements: co-evolution or useful parasites? *Heredity* 2000;**85**:101–6.
14. Slotkin RK, Martienssen R. Transposable elements and the epigenetic regulation of the genome. *Nat Rev Gen* 2007;**8**:272–85.
15. Feschotte C. The contribution of transposable elements to the evolution of regulatory networks. *Nat Rev Genet* 2008;**9**:397–405.
16. Ross ED. Modulation of affect and non-verbal communication by the right hemisphere. In: Mesulam M, editor. *Principles of behavioral neurology*. Philadelphia: F. A. Davis; 1985. p. 239–53.

17. For a recent special issue on brain laterality see Tzourio-Mazoyer N, Seghier M, editors. *The neural basis of hemispheric specialization*. *Neuropsychologia*, vol. 93B. 2016. p. 319–512.
18. Nisbett R. *The geography of thought: how Asians and Westerners think differently...and why?*. New York: The Free Press; 2003.
19. Skiyokawa K, Dienes Z, Tanaka D, Yamada A, Crowe L. Cross Cultural differences in unconscious knowledge. *Cogn* 2012;**124**:16–24.
20. Ivry R, Robertson LC. *The two side of perception*. Cambridge (MA): MIT Press; 1998.
21. For support for this view with natural scenes see Peyrin C, Chauvin A, Chokron S, Marendaz C. Hemispheric specialization for special frequency processing in the analysis of natural scenes. *Brain Cogn* 2003;**53**:278–82.
22. Chen Q, Marshall JC, Weidner R, Fink GR. Zooming in and zooming out of the attentional focus: an fMRI study. *Cereb Cortex* 2009;**19**:805–19.
23. Fink GR, Marshall JC, Halligan PW, Frith CD, Frackowiak RF, Doland RJ. Hemispheric specialization for global and local processing: the effect of stimulus category. *Proc R Soc Lond B* 1997;**264**:487–94.
24. Kosslyn SM, Chabris CF, Laeng B. Asymmetries in encoding spatial relationships. In: Hugdahl H, Davidson R, editors. *Brain asymmetry*. Cambridge (MA): MIT Press; 2004. p. 303–39.
25. Laeng B, Zarrinpar A, Kosslyn SM. Do separate processes identify objects as exemplars versus members of basic-level categories? Evidence from hemispheric specialization. *Brain Cogn* 2003;**53**:15–27.
26. For a discussion of some of the complexities of the association between of complexities handedness and hemispheric dominance see Springer S, Deutsch D. *Left brain right brain: perspectives from cognitive neuroscience*. New York: W. H. Freeman; 1998. p. 131.
27. Miller G. The magical number seven plus two: some limits on our capacity for processing information. *Psychol Rev* 1956;**63**:81–97.
28. Putnam H. *Reason, truth and history*. Cambridge (UK): Cambridge University Press; 1981. p. 50.
29. Gazzaniga MS, Le Doux JE. *The integrated mind*. New York: Plenum Press; 1978.
30. Gazzaniga MS. Cerebral specialization and interhemispheric communication: does the corpus callosum enable the human condition? *Brain* 2000;**123**:1293–326.
31. For a review of some of the evidence showing that familiar patterns may increasingly activate the left hemisphere see Goldberg E. *The wisdom paradox: how your mind can grow stronger as your brain grows older*. New York: Gotham Books; 2005.
32. Goël S. Fractionating the system of deductive reasoning. In: Berntson GG, Cacioppo JT, editors. *Handbook of neuroscience for the behavioral sciences*, vol. 1. Hoboken (NJ): Wiley; 2009. p. 426.
33. Goël V, Tierney M, Sheesley L, Bartolo A, Vartanian O, Grafman J. Hemispheric specialization in human prefrontal cortex for resolving certain and uncertain inferences. *Cereb Cortex* 2007;**17**:2245–50.
34. Goël V, Stollstorff M, Nakic M, Knutson K, Grafman J. A role for right ventrolateral prefrontal cortex in reasoning about indeterminate relations. *Neuropsychologia* 2009;**47**:2790–7.
35. Goël V, Grafman J. The Role of the right prefrontal cortex in ill-structured problem solving. *Cogn Neuropsychol* 2000;**17**:415–36.
36. Goël V, Vartanian O. Dissociating the roles of right ventral lateral and dorsal lateral prefrontal cortex in generation and maintenance of hypotheses in set-shift problems. *Cereb Cortex* 2005;**15**:1170–7.
37. Beeman MJ, Bowden EM. The right hemisphere maintains solution-related activation for yet-to-be solved insight problems. *Mem Cogn* 2000;**28**:1231–41.
38. Mednick SA. The associative basis of the creative process. *Psychol Rev* 1962;**69**:220–32.
39. Jung-Beeman M, Bowden EM, Haberman J, Frymiare JL, Aerameel-Liu S, et al. Neural activity when people solve verbal problems with insight. *PLoS Bio* 2004;**2**:0500–10. https://doi.org/10.1371/journal.pbio.0020097.

## APPENDIX 1

Our previous study showed that only specific small breakdown products were released from intact cells that had incorporated radiolabeled RNase, which was taken up into their lysosomes.[1] In this experiment, we wanted to confirm that similar peptides would be released from lysosomes loaded with radiolabeled RNase that were isolated from the cell and incubated at either 0°C or 24°C. for 2h.[2] All samples were run over a molecular sizing column to separate the various low–molecular weight breakdown fragments. The X-axis in Fig. A1.1, labeled *relative mobility*, corresponds to the fraction in which the radiolabeled fragments eluted from the column. Relative mobility is proportional to the molecular weight of the fragment under the specific conditions used. Relative mobility increases from left to right because the smaller fragments move more slowly through the column.

We were especially interested in the most plentiful small fragment that is released from the intact cells, a peptide running at a relative mobility of 0.6 (molecular weight = 400). The graphs suggest that the 0.6 relative mobility fragment was indeed released from the isolated lysosomes, but surprisingly only at the higher of the two temperatures. Even more unexpectedly, this fragment was not detectable within the lysosomes at either temperature. Instead, almost all radioactivity remaining within the organelle at both temperatures was found in a slightly larger fragment not seen before, running at a relative mobility of 0.5 (molecular weight = 550). The final surprise was that only small amounts of this larger fragment were released into the bathing medium at either temperature (Fig. A1.1).

My colleagues questioned whether the data demonstrated that the small peptide released from the isolated lysosomes at 24°C corresponded to the same peptide released by the intact cells. They pointed to the fact that although the peak in the graph of material released from lysosomes at 24°C was centered at a relative mobility of 0.6, it was considerably broadened, which suggested the possibility that it corresponded to other peptides instead of, or in addition, to, the peptide of interest. If I could not be certain that the peak represented the smaller peptide, how could I say that the release was selective?

Yet when all the data are considered together, several internal controls support the validity of the conclusions. Most important is the marked *absence* of any radiolabel at a relative mobility of 0.6 in the material released from lysosomes at 0°C. Its absence delineates a clear trough centered on a relative mobility of 0.6. Thus at 0°C, little or none of the smaller peptide is released

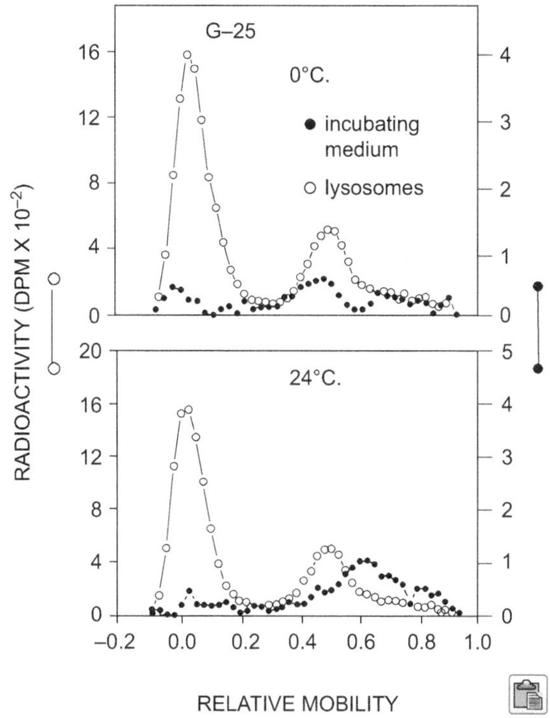


**Figure A1.1** Isolated lysosomes loaded with radiolabeled RNase were incubated at either 0°C or 24°C. for 2 h and then separated from their incubating medium. All samples were run over a column sizing column that separates out low-molecular weight fragments to determine which fragments were retained by the lysosomes and which released into the incubating medium at each temperature. *DPM*, disintegrations per minute. *(Reproduced from Isenman LD, Dice JD. Secretion of intact proteins and peptide fragments by lysosomal pathways of protein degradation.* J Biol Chem *1989;**264**:21591–6. Reprinted by permission from Elsevier.)*

even though at both temperatures roughly the same amount of the larger peptide is released.

Nonetheless, that the broadened peak of released peptides centered at a relative mobility of 0.6 at the higher temperature suggests that other molecules are also released. However the peak height at a relative mobility of 0.6 is roughly twice what it is at a relative mobility of 0.5. This is consistent with the claim of selective release across the lysosomal membrane, because it suggests that twice as much of the smaller peptide as the larger is released at this temperature. Importantly, this is true even though the larger molecule is the most abundant molecule within the lysosomes. The amount of the larger peptide released at either temperature reflects less than 10% of its total amount in the system (note the scaling change between the lysosomes and the incubating medium). In contrast, almost *all* of the smaller peptide in the system is released at the higher temperature, because little or none is found within the lysosomes at either temperature. (Most of the radiolabel remaining within the lysosomes at a relative mobility of 0.6 can be accounted for the larger and smaller flanking peptides.)

Another way to demonstrate that the smaller peptide represents the major molecule released at the higher temperature is to subtract the graph of material released at 0°C from the graph of material released at 24°C. Fig. A1.2 shows the calculated difference (along with the superimposed data for release at 24°C) and reveals that the peak is narrowly centered over a relative mobility of 0.6, with roughly 85% of the material released falling in this peak. This suggests that substantial amounts of the 0.6 relative mobility peptide are released at the

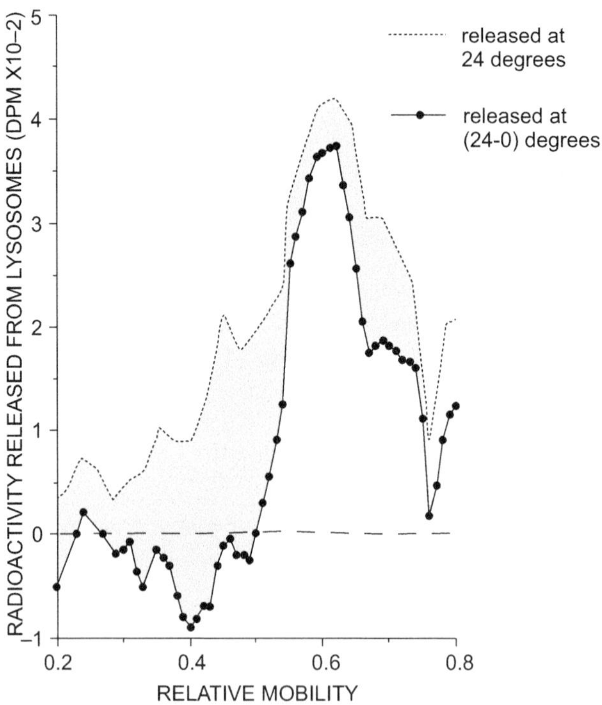


**Figure A1.2** Subtracting the radioactivity released into the incubating medium at 0°C. from the radioactivity released into the incubation medium at 24°C reveals that the difference largely results from an increase in the fragment that falls at a relative mobility of 0.6. *DPM*, disintegrations per minute.

higher temperature and that the predominant response to the increase in temperature is the release of this peptide. There is a smaller peak at roughly a relative mobility of 0.7, which may correspond to an alternative form of the 0.6 relative mobility peptide, or perhaps to a distinct peptide released at the higher temperature. However, much of the radioactivity in this peak represents the falling phase of the large peak at a relative mobility of 0.6.

## APPENDIX 2

### Different Levels to the Conclusion

1. Peptide transport occurs across the lysosomal membrane.
2. Peptide release is selective.
3. Release of the smaller peptide is temperature dependent; however, its formation rather than its transport may be the rate-limiting factor.
4. Release of the peptide occurs rapidly because little remains in the lysosomes.

### An Attempt at a Linear Exposition of the Evidence

These conclusions, along with the evidence backing each of them, illustrate the interwoven nature of the different levels of the observations. The subsequent analysis demonstrates that because of this interweaving, any attempt to understand the conclusions in a linear or systematic fashion necessarily involves considerable redundancy.

### The Evidence for Each Level of the Conclusion

1. Peptide transport occurs across the lysosomal membrane.
   Conclusion 2 (peptide release is selective), Conclusion 3 (release of the smaller peptide is temperature dependent; however, its formation rather than its transport may be the rate-limiting factor), and Conclusion 4 (release of the peptide occurs rapidly because little remains in the lysosomes) all provide internal evidence supporting the view that the release of peptides across the lysosomal membrane is real and not an artifact, resulting from broken or leaky lysosomes.
2. Peptide release is selective.
   Conclusion 3 (release of the smaller peptide is temperature dependent; however, its formation rather than its transport may be the rate-limiting factor): Only the release of the smaller peptide changes substantially with the rise in temperature, as shown by the difference between what is released at the higher and lower temperatures (Fig. A1.2).

Conclusion 4 (release of the peptide occurs rapidly because little remains in the lysosomes): At both temperatures, most of the material remaining in lysosomes at a relative mobility of 0.6 can be accounted for by larger and smaller flanking peptides.

At the higher temperature, relatively little of the larger peptide, the most abundant low–molecular weight peptide in the lysosomes, is released into the incubating medium, whereas almost all of the smaller peptide is released.

In addition, when three synthetic radiolabeled peptides from RNase of a similar isoelectric point to the small natural degradation fragment that were loaded into isolated lysosomes, the ratio of secreted to retained peptide varied over almost a fivefold range.[2]

3. Release of the smaller peptide is temperature dependent; however, its formation rather than its transport may be the rate-limiting factor.
   Conclusion 4 (release of the small peptide occurs rapidly, because little remains in the lysosomes): At both temperatures, much of the material remaining in lysosomes at the position of the smaller peptide seems to reflect both a larger and a smaller flanking peptide. Thus, it is likely that little of the smaller peptide is made at the lower temperature. If this is true, the enhanced release at the higher temperature may depend more on an increased rate of formation of the peptide rather than on an increased *rate* of release.

   None of the smaller molecule is released at the lower temperature, but a substantial amount is at the higher temperature, as shown by the difference between what is found in the medium at the higher and lower temperatures (Fig. A1.2). This graph also indicates that only release of the smaller molecule and not the release of the larger molecules is affected by the enhanced temperature to any substantial degree.
4. Release of the peptide occurs rapidly once it is formed, because little remains in the lysosomes.
   The rate of release at the higher temperature allowed most if not all of the smaller peptide to appear in the incubating medium.

   Even if the change in temperature had no effect on the rate of release of the small peptide, but only its formation, the conclusion that peptide release from lysosomes is selective still holds for the reasons given in the second and third paragraphs under 2 (peptide release is selective).

## REFERENCES

1. Isenman LD, Dice JD. Secretion of intact proteins and peptide fragments by lysosomal pathways of protein degradation. *J Biol Chem* 1989;**264**:21591–6.
2. Isenman LD, Dice JD. Selective release of peptides from lysosomes. *J Biol Chem* 1993;**268**:23856–9.

# CHAPTER 4
# The Brain and Perception

## Contents



The parallel-interconnected information processing at the core of intuitive cognition mimics the connectivity of the nervous system. Neurons, the primary cells in the nervous system, form connections with many other neurons. Those that tend to fire at the same time become more strongly connected to each other. Eventually they form networks that can process together different features of an object, or different objects that tend to be associated with each other. This parallel-interactive connectivity of the brain accounts for its *plasticity*, or its ability to become automatically molded by experience. This in turn explains many of the characteristics of unconscious intelligence and intuition.

## CLASSICAL VERSUS NATURAL VIEW OF INTELLIGENCE

At the most elementary level, all mental functioning necessarily depends on neurons and their parallel-interactive connections. However large numbers of neurons can become organized in such a way that they produce the rational, stepwise processing that is generally associated with conscious mental function. In the early days of artificial intelligence, most researchers thought this higher level must be the most important one for human intelligence. Such a perspective reflects the traditional view of intelligence that focuses primarily on its *symbolic* aspects: the manipulation of verbal and mathematical symbols in a logical, or rule-based, manner.

According to this view, intelligence is a closed formal system like mathematics. At its most basic level of description, it is something apart from the


*Understanding Intuition*
ISBN 978-0-12-814108-3
https://doi.org/10.1016/B978-0-12-814108-3.00004-X

biological characteristics of the body and brain that house it as well as from the environment with which it interfaces, i.e., apart from the physical world. This kind of top-down, information-processing architecture underlies the functioning of digital computers.

A computer's central processing unit, a control unit analogous to an idealized view of the rational conscious mind, performs logical operations stepwise on data stored in its memory banks. It is a universal computing machine in the sense that all objects and events are treated as symbols, and the logical operations these symbols undergo are in all cases the same. Information is presented to the machine as a series of digits generally represented as strings of 0 and 1. What the data corresponds to and how it is acquired is unimportant. This view of intelligence, which was prevalent throughout much of the 20th century, underpinned the attempt of many cognitive scientists to understand how the mind works without taking the physical brain or the body to which it is connected into account.

At the core of all intelligence is the ability to perceive similarity and difference. We perceive *this* to be similar or dissimilar to *that*, so we respond to it in either the same or in a different way. At one level, the conflict between the older, traditional, top-down, central processing view of intelligence and the one that fits with unconscious processing and intuition is: How do basic categories arise that we use to break the world up into "kinds"?

According to the classical view, categories are analogous to mathematical symbols. They are composed of objects or events that satisfy *all* the conditions that define the group. Although the members of a category may differ in an infinite number of other dimensions, they share all of the characteristics that define the category.

Category formation in the type of unconscious intelligence that characterizes much of animal and human behavior is a much messier affair. Rather than the presence of necessary and sufficient characteristics, it depends on the way a variety of often vaguely defined, or diffuse, characteristics come together below awareness to form a pattern. Natural or intuitive categories tend to form around prototypes, central, or perhaps ideal, examples that best characterize the category based on experience. By the same token, members of natural categories are graded in strength.[1] For instance, a robin is perceived as a more central member of the category *bird* than an ostrich, and some shades of red are perceived as more central members of the category red than others.[2]

In children, natural categories tend to form first around objects that fall in a middle range of specificity. For example, chair is a basic level category, as opposed to either *furniture*, a higher-level category, or to *kitchen chair*, a lower-level category. Adults also recognized basic level objects more quickly than other levels.[2]

Moreover categories often correspond to a mode of physical interaction.[3] What characterizes chairs is that we sit on them, generally with bended knees, and they most often provide some support for the back. Intelligence according to this more natural perspective is a *bottom-up* process grounded in the interaction of the biological organism and its environment.

Each category is recognized according to a general gestalt that is built up below awareness over time from various examples. We can recognize a big cat as a cat and a tiny dog as a dog without giving it any thought, yet it is difficult to characterize the difference between them rigorously. These general gestalts can encompass a variety of different finer-grained patterns: both cats and dogs come in a variety of breeds as well as mixtures of breeds.

Furthermore, we can frequently recognize the relevant pattern even with partial information. We can often distinguish between a big cat and a small dog—or between a Collie and a German Shepherd—if only its face or perhaps only a portion of it, or only its body or a portion of it is visible. This is of great importance because objects are often partially hidden by others, seen in poor light, or at a distance. Instead of the presence of certain necessary and sufficient characteristics, natural judgments about categories are based on the interaction of multiple soft, or flexible, constraints to form recognizable patterns.

Clinical studies of brain-damaged patients support the view that the brain does not function like a top-down processor. If it did, damage to its central processing unit should have a drastic effect on *general* intelligence. Yet localized damage to *higher brain areas* tends to cause specific rather than general defects in all-over intelligent functioning. General intelligence decays in a gradual, or a "graceful" way, with generalized brain cell loss, as may occur with aging.[4]

In contrast, connectionist or neural network information processing architecture, which is modeled on the way neurons connect to each other, can readily account for natural categorization and graceful degradation, along with many other aspects of human intelligence. It depends on what is sometimes referred to as the *subsymbolic* aspects of intelligence, the distributed patterns of often vague features that converge below awareness to determine the recognition of objects and events.[5] Neural network processing excels in situations that require multiple component parts to be perceived in the context of each other to make a pattern, such as recognizing faces and deciphering handwriting.

## ARTIFICIAL NEURAL NETWORKS

Artificial neural networks can readily reproduce many aspects of animal and human intelligence that tend to occur below awareness. Moreover neural network architecture is easy to replicate on digital computers. Artificial neural networks are composed of rows of units, which are analogous to neurons, and the connections between units, which are analogous to the synapses connecting neurons. When one of the unit fires, each of the other units connected to it might also fire, depending on the strength of the connection.

At a minimum, a network consists of a row of input units and a row of output units. Each unit in the input layer is usually connected to every unit in the output layer. Like the brain networks on which they are modeled, neural networks after they have been trained can process a variety of different features and their interactions at the same time. To "train" an artificial network means to adjust the strength of the connections between the different units (called *weights*) using a large number of known examples of each of the different categories the network will be required to distinguish. Training is complete when activating the specific input pattern associated with each sample automatically activates the appropriate output unit—for instance, that associated with cat as opposed to dog or raccoon.

Many neural networks have an additional layer of units between the input layer and the output layer called *hidden units*. They allow the network to extract from the sample set during training the common features or the common relationships that facilitate the transformation of the input patterns into the appropriate outputs. The additional layer of units increases the number and complexity of patterns a network can encode.

Hidden units can be used to encode classical categories. Each hidden unit can come to represent a specific category; it becomes maximally activated when the input consists of all the features that correspond to the specific category. However hidden units are even more useful for natural categories, which, as the last section suggested, are not so cut-and-dried. Hidden units come to represent the statistical co-occurrence of certain features during training. Therefore they can encode prototypes, or the subtle interweave of features that best characterizes a natural category.[6] They also allow networks to come up with the best match for less prototypical, or less central, members of a category.

In addition, input features, or characteristics, are most often a matter of degree, such as height, redness, or hairiness. Likewise, input units can be activated to different extents and indicate the relative strength of the different characteristics they happen to represent.[7] These graded profiles are sometimes called *vectors*. Graded encodings also increase the number of patterns a network can discriminate.

Each hidden unit, like each input unit, can also be activated to a variety of different levels. The output units may then come to respond to different patterns of activation across the whole hidden unit layer, i.e., which hidden units are activated and to what extent. The possibility of different patterns of activation of the hidden unit layer also greatly enhances the ability of a neural network to make difficult discriminations.

Artificial neural networks developed to distinguish between mines and rocks on the ocean floor illustrate the potential power of networks with graded encoding capacity in their input and hidden units.[6,7] Sonar probes of the ocean bottom pick up rocks as well as mines, but the two cannot be differentiated by any single sonar frequency or fixed combinations of frequencies. Neural networks are ideally suited to separating them based on of their sonar echo profiles, or "fingerprints."

One such network is shown in Fig. 4.1. It consists of 13 input units, each of which connected to seven hidden units. Each hidden unit is in turn connected to two output units, one representing rocks and the other mines.

The training set consisted of the sonar echo recordings of a large series of different rocks and different mines that had been analyzed at 13 separate sonar frequencies. The 13 frequency values for each sample (normalized between 0 and 1) were fed into the appropriate frequency input units. The overall goal of training was to adjust the weights, or the strength of the connections between units, so that each echo profile or fingerprint of a mine as much as possible activated only the mine output unit, and each echo profile of a rock as much as possible activated only the rock output unit. Initially all of the weights were set at random.

A computer-run back-propagation algorithm functioned as the "teacher." After each sample was entered, the back-propagation algorithm compared the actual response of the network with the desired response and examined which weight changes would make the desired output occur. It then nudged each of these weights a little bit in the indicated direction. It moved each one only a little because the goal of training was not to make

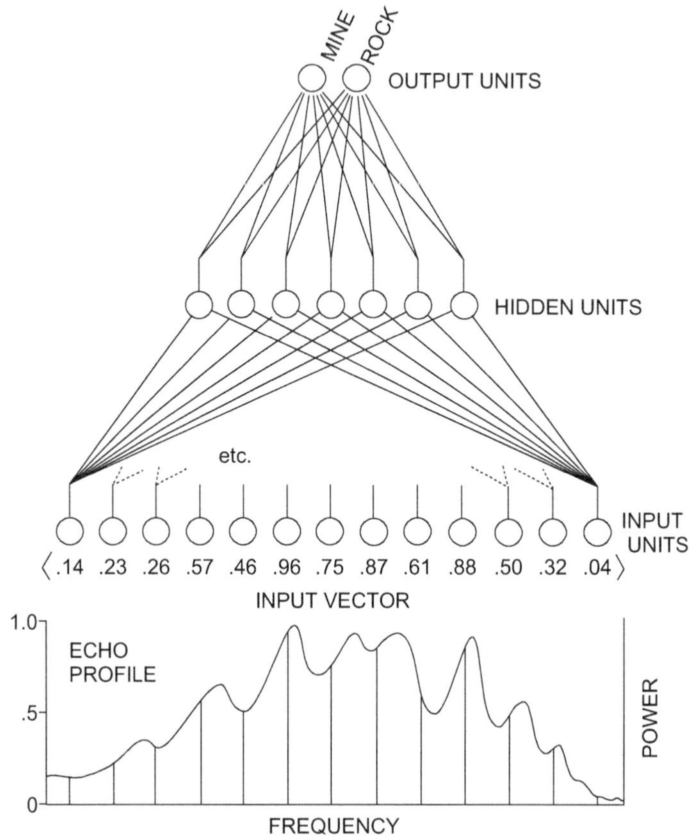


**Figure 4.1** ***Neural network for distinguishing rock and mines on the ocean floor based on their sonar echo profiles.*** The network consisted of 13 input units, each of which was connected to seven hidden units, which were in turn connected to two output units, one for rocks and one for mines. For training, the sonar echos from 50 different rocks and 50 different mines that had been analyzed at 13 separate frequencies to generate an echo profile, (the power at each frequency represented as a fraction of one) were fed into the appropriate frequency input unit. Together these thirteen values form an input vector. *(Reproduced from Churchland P.* A neurocomputational perspective. *Cambridge (MA): MIT Press; 1992. p. 203. Reprinted with permission of MIT Press.)*

only this one reading register correctly, but rather to accommodate a wide variety of different examples with one final series of weight settings. To accomplish this, it was necessary to enter the training set many times.

The sonar network was able to distinguish the training set completely. This means that the patterns of activation across the hidden unit layer (the patterns formed by the strength of activation of its different hidden units) came to have profiles that corresponded to prototypical examples of mines and to prototypical examples of rocks. As Paul Churchland pointed out, the activation pattern across the seven hidden units for each sample after training defined a unique point in an abstract seven-dimensional space.

He wrote, "what the system is searching for during the training period is a set of weights that *partitions* this space so that any mine input produces an activation vector across the hidden units that falls somewhere within one large subvolume of this abstract space, while any rock input produces a vector that falls somewhere into the compliment of that subvolume."[6(p168)] Prototypical samples fall near the center of their subregion whereas more problematic or ambiguous samples fall toward the edges. A network with three or seven possible outcomes, for example, would search for a set of weights so that the activation patterns across the hidden-unit layers partitioned the space into three or seven subregions.

After the *training phase*, the weights between units were fixed. During the subsequent *running phase* when the network was challenged with unknown samples, it was 90% accurate in telling rocks from mines. This success rate was better than for experienced human sonar operators performing the same task.

The ability of the network to make the required distinctions was not a foregone conclusion. Rather it depended on the presence in the training samples of representative patterns that distinguish rocks from mines and, moreover, patterns that could be captured by the structure of this particular network. Similarly, there was no way to know in advance what the individual hidden units would come to encode: that is, the input unit patterns that would activate any of the individual hidden units maximally, or, more important, the activation patterns across the entire hidden unit layer that would activate either the rock or mine output unit maximally. The process developed on its own during the training period, and, again, it depended on the training set and the structure of the network.

Computer analysis of the network after training revealed the regularities its hidden units had learned to encode. It turned out that the final division of the hidden-unit hyperspace into rock and mine subregions by their patterns of activation corresponded to the way metal and nonmetal sonar targets would partition the space. Although during training the network was given only the sonar echo profiles of samples as input, its hidden units extracted the natural category metal.

The ability of humans to recognize a large number of faces under a variety of obscuring conditions has been honed by evolution to an especially

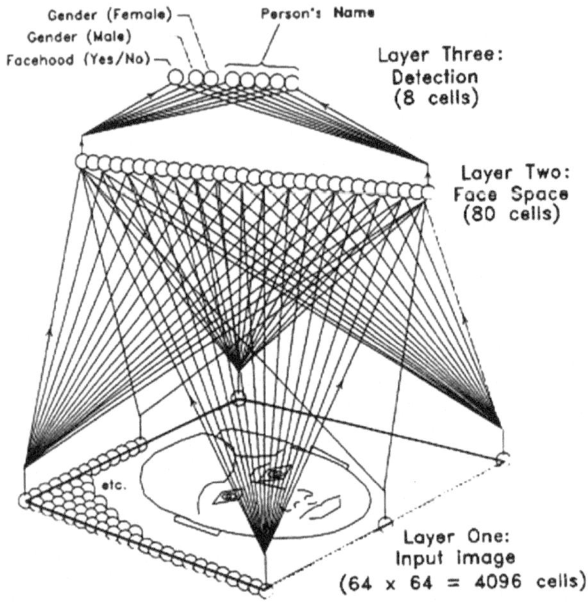


**Figure 4.2** ***Example of an early face recognition network.*** The network consisted of a 64×64-input unit grid, with each unit sensitive to 256 degrees of brightness between black and white: 80 hidden units and 8 output units. For training, the researchers used a set of 64 photos, which included different shots of 11 faces as well as a number of nonfaces. The network learned to discriminate faces from nonfaces and males from females, as well as novel pictures of individual familiar faces. *(From Churchland P.* The engine of reason, the seat of the soul. *MIT Press; 1996. p. 40–41, based on the work of Gary Cottrell. Reprinted with permission from MIT Press.)*

fine degree. It is so central to survival that a special region of the brain is largely dedicated to it. Babies are able to discriminate faces very early, which suggests that the perceptual networks responsible are either prewired or particularly easily molded by experience. Computers using traditional top-down symbolic approaches were initially unable to recognize faces, yet simple artificial neural networks easily did so.

Fig. 4.2 illustrates an early face recognition network.[8] It consists of a 64× 64-input unit grid, with each unit sensitive to 256 degrees of brightness between black and white, 80 hidden units, and 8 output units. For training, the researchers used a set of 64 photos that included different shots of 11 faces as well as a number of nonfaces (Fig. 4.3). The network learned to discriminate faces from nonfaces, males from females, as well as the individual faces.

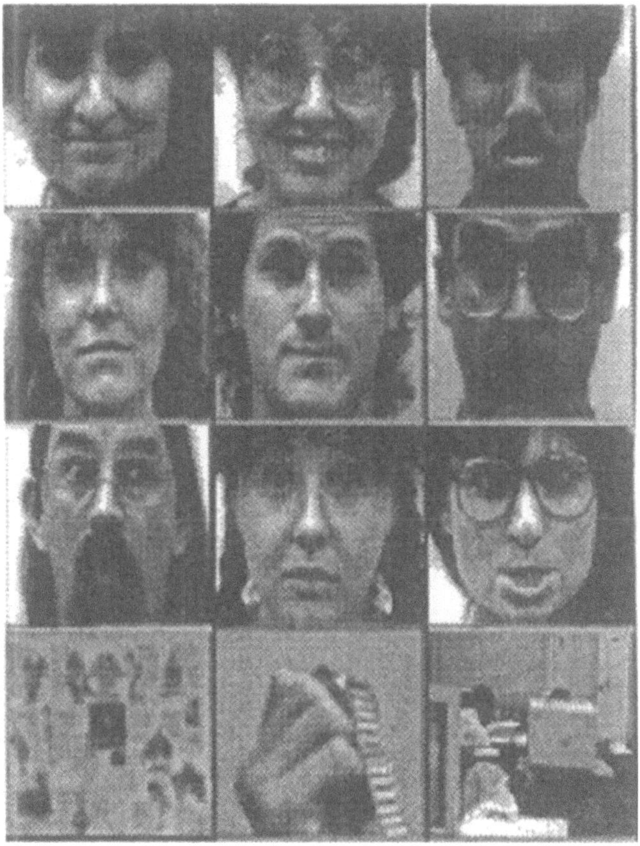

**Figure 4.3** ***Some photos of faces as well as objects used to train the face recognition network.*** The network was trained on 64 photos from 11 faces and 13 nonface photos. *(From Cottrell GW. Extracting features from faces using compression networks: Face, identity, emotion, and gender recognition using holons. In: Touretzky S, Elman J, Sejnowsky TJ, editor.* Connectionist models: proceedings of the 1990 summer school. *San Mateo (CA): Morgan Kaufmann; 1991. Reprinted with permission from Elsevier.)*

After training, analysis of the network revealed a surprise. Different specific hidden units had *not* as would seem likely become dedicated to representing various features, such as the eyes, nose, and mouth, or to representing important relationships, such as the distance between the eyes. Rather each hidden unit had become maximally activated in response to a slightly different construction of a *face as a whole* (Fig. 4.4). These widely distributed aspects of a face recognized by each hidden unit were named *holons*.

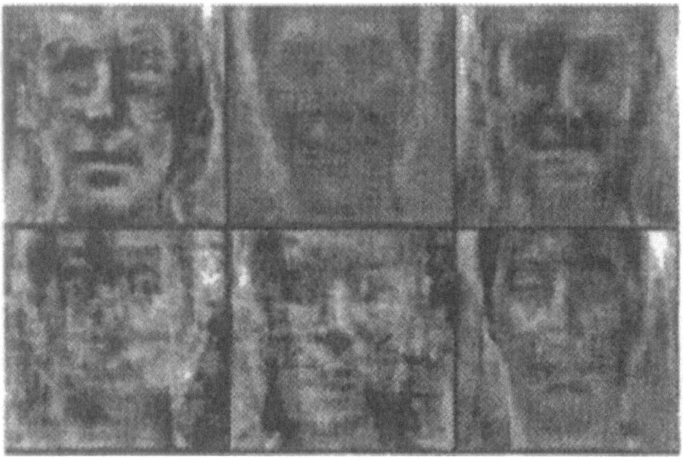

**Figure 4.4** ***What the hidden units of the face recognition network encode.*** Preferred stimuli of several of the hidden units after training. Each hidden unit came to encode the face as a whole. Although they are full faces, they are not specific to any of the individual faces. *(From Cottrell GW. Extracting features from faces using compression networks: Face, identity, emotion, and gender recognition using holons. In: Touretzky S, Elman J, Sejnowsky TJ, editor.* Connectionist models: proceedings of the 1990 summer school. *San Mateo (CA): Morgan Kaufmann; 1991. Reprinted with permission from Elsevier.)*

Recognizing a given face depended on the pattern of activation across the whole hidden-unit level. For example, in response to a previously unseen picture of one of the faces, many of the 80 hidden units were close to maximally activated while many others were hardly at all. Moreover the researchers found that the network was able to recognize a face even if a wide band hid the eyes and the top of the nose (Fig. 4.5). Because each of its hidden units was sensitive to the face as a whole, the network could fill in the missing information with a surprising degree of accuracy based on the other information. This does not imply that human face recognition actually works in this way. Rather it illustrates how a network with no special prewiring might organize itself to recognize a number of different faces and their genders.

The ability to recognize patterns that change in time, as well as to reproduce a wide variety of other human behaviors, such as generating a sequence of movements, can be incorporated into artificial neural networks by adding feedback. One such network, called a *simple recursive network*, has a loop that feeds the pattern of activation that developed across the hidden unit layer from the previous input back into the network with the current input.[9]

**Figure 4.5** ***Recovery of missing portion of a face.*** When the trained network was given a new photo of a familiar face with a bar obscuring 20% of the face, it was able to fill in the missing information. *(From Cottrell GW. Extracting features from faces using compression networks: Face, identity, emotion, and gender recognition using holons. In: Touretzky S, Elman J, Sejnowsky TJ, editor.* Connectionist models: proceedings of the 1990 summer school. *San Mateo (CA): Morgan Kaufmann; 1991. Reprinted with permission from Elsevier.)*

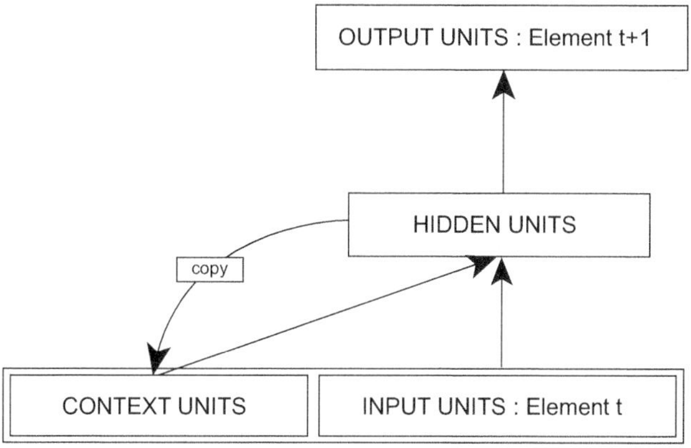


**Figure 4.6** ***A simple recurrent network.*** Context units feed the pattern of activation that developed over the hidden units for the prior element back into the input for the current element. This provides the input units with information about the features that were essential to optimizing output for the previous element. In such a fashion, the patterns of activation over the hidden units can come to encode relevant information about successively more elements in a sequence. *(From Cleeremans A, McClelland J. Learning the structure of event sequences.* J Exp Psychol Gen *1991;**120**:235–53.)*

These *context units* incorporate what turned out to be essential to the network's encoding of the prior input into the current input, eventually generating sequences. Fig. 4.6 shows a simple recursive network that mimics the ability of human subjects to learn without effort, or the awareness of learning anything, nonrepeating sequences of numbers generated by computer from an artificial grammar, as described in Chapter 2.[10]

## THE BRAIN AS A NONLINEAR DYNAMIC SYSTEM

Artificial neural networks with their feature detectors and automatic output responses are pattern recognition machines. They can mimic and enhance many aspects of human sensory recording as well as some of the predictive capacity of intuition. They are becoming ever more useful.[11] However, to begin to account for the novel and astoundingly rich mental integration possible with human intuition, a more dynamic model of brain function is required. It must be able to account for ongoing changes in the *internal meaning* of objects and events, as well as changes in the external world.

Real neurons in association can take advantage of the many additional levels of connectivity in the brain. Most neurons receive feedback from multiple other layers and regions, which is critical to brain function. In the human visual system, for example, feedback pathways greatly outnumber feed-forward pathways.[12] They can also connect to other neurons in their layer. However, in contrast to units in artificial networks, real neurons tend to be only relatively sparsely connected to neurons in the next higher level.[13]

Most of the early data on sensory processing came from studies using anesthetized animals responding to greatly simplified stimuli. Simple artificial neural networks proved especially useful in simulating the results of these studies, helping to provide a basic level of understanding of how the brain processes sensory information.[14] However anesthesia attenuates the feedback connections from higher to lower levels in neural tissue as well as the lateral connections between neighboring neurons within a tissue layer.[15] These additional levels of connectivity make even basic sensory recording much more dynamic and complex than earlier studies indicated. Many studies now use alert animals and more lifelike sensory stimuli.[16]

Moreover perception and behavior are directed by much more than what the sensory organs happen to transmit about the environment. Until relatively recently, most brain imaging studies filtered out the internal activity of the brain, either by time-averaging responses or by using baselines that subtract activity other than that directly *evoked* by the stimulus. These *evoked responses* account for only a small amount of brain activity, since they increase the brain's oxygen use by only about 5%. More researchers are now examining the brain's internal intrinsic activity, part of what is sometimes called its "dark energy."[17]

As early as the 1970s, neurobiologist Walter Freeman began to explore the implications of the brain's internal activity for perception and behavior. As he put it:

> *...[R]esearchers spend, as they should have spent, small fortunes on the care, feeding, and housing of their subjects; they tailor the equipment and tasks to the capabilities of the species, familiarize and train them, and then sit waiting for them to deign to stop eating, licking, grooming, or just looking around long enough for the experimenter to get in a CS [conditioned stimulus] for a controlled trial; all this can go on for weeks. What is lost in all this is the fact that these animals are continually producing behaviors from within by anticipating external stimuli to guide or pace their actions. These behaviors express internally generated activity of the nervous system and are not deterministic responses to stimuli.*[18]

Freeman's work provides a complex and integrated account of brain function based on nonlinear dynamical systems theory that can begin to grapple with the behaving organism. It suggests a mechanism by which an animal's "inner life"—or at least the complex neural activity of its brain—governs what it perceives, and thus to a large extent governs its behavior. Freeman's understanding of sensory perception in lower animals lays the foundation for understanding how the novel higher-level perception below awareness that is at the core intuition might come about. His experiments argue for a biologically tethered sphere of mind that shapes perception below the level of awareness. It is based on the *meaning* of objects and events as they relate to immediate and long-term goals and to the sum of the knowledge and experience of individual animals.

## Perception and Complexity in the Brain

The brain must filter out unimportant incoming sensory information at the same time that it intensifies what is important. Freeman became interested in what *perception*, rather than sensory recording alone, might mean at the neural level, as captured by the electroencephalogram (EEG). He initially studied olfaction in small, awake mammals such as cats and rabbits to learn, as the title of one of his books puts it, *How Brains Make Up Their Minds*.[13]

In neural tissue, the axon of one neuron and the dendrite of another are separated by small gaps called *synapses*. When an axon fires, neurotransmitter molecules rapidly diffuse across the synapse and bind to receptors on the opposing neuron's dendrite. This initiates a wave of electrochemical excitation that travels along the dendrite to the opposing neuron's cell body. In turn, this generates a weak electrical current in the extracellular fluid outside the dendritic membrane. The currents in the extracellular fluid from different cells add together. When summed, they can greatly enhance the possibility that other neurons in the neighborhood will also fire. The brain waves recorded by EEG primarily measure the combined extracellular current generated by the dendrites of a large number of cells.[13]

Extracellular currents become particularly important in the presence of neuromodulator molecules that signify biological "value." These values correspond to the animal's intrinsic needs and goals, and they are progressively fine-tuned by experience. Neuromodulators work by modifying neural firing to encourage behaviors that support survival and well-being. They are released in response to signals from the limbic system, the group of brain structures largely responsible for innate seeking and avoidance behaviors and for emotional responses. Some neuromodulators such as dopamine and acetylcholine (sometimes called nonspecific neuromodulators or neurotransmitters) are released throughout large regions of the brain. Others are deployed more strategically in both the brain and body at the behest of the emotional and motivation system. Oxytocin, for example, is released in more specific regions of the brain and body during childbirth, sexuality, and bonding.[19]

In the presence of neuromodulators, nearby neurons that fire together readily form associations that are strongly interconnected, called *cell assemblies*. When one neuron in a cell assembly fires, others in the assembly are more likely to fire as well. The large extracellular current that results when a cell assembly fires enhances the probability that neighboring neurons will also fire.[13] Moreover, because of positive feedback, the more cells that fire, the more likely it is that other cells in the layer will also fire (up to a point). Together they can spark a response of the tissue layer as a whole. The individual cells undergo a *state or phase transition*; they begin to function as a coordinated population, which can be detected by EEG.

## Chaos and the Brain

A bit of background about complex dynamical systems is helpful for understanding Freeman's experimental work. Complex systems such as the brain are composed of autonomous but interconnected parts that feed back on each other. The extensive connections between higher and lower layers in the brain, along with the potential lateral connections between cells in individual layers, make each of the subsystems as well as the entire brain highly interactive. As a consequence, a small local event can sometimes cause large-scale and even global changes in organization. Complex systems are capable of *self-organization*, a reorganization potentially sparked by a phase transition in one layer of one component that drives the whole system into a different and more highly organized state.

A story about the 17th-century physicist and clockmaker Christiaan Huygens illustrates the basic idea behind self-organization using a simple one-layer, laterally interconnected system.[20] Huygens noticed that two pendulum clocks he had built were swinging in lockstep on a wall outside his shop. Moreover, whenever he tried to disturb their synchrony, they eventually came back into phase. After a while he realized that weak vibrations between them were traveling in the wall and coupling, or entraining, their activity.

The behavior of nonlinear interconnected systems over time is governed by *attractors*, functions that describe how the system settles down, or the behavior the system is "attracted to" over time, given a wide variety of different starting conditions. Huygens' clocks on the wall that came into phase without regard to when they were started are an example of a nonlinear interconnected system with a *point* attractor, one of three types of attractors (Fig. 4.7). Systems with point attractors, such as a pebble in a cone, always end up in the same place. Drop it in the center or anywhere down the sides, and it will end up on the bottom. The walls of the cone define its *basin of attraction*.

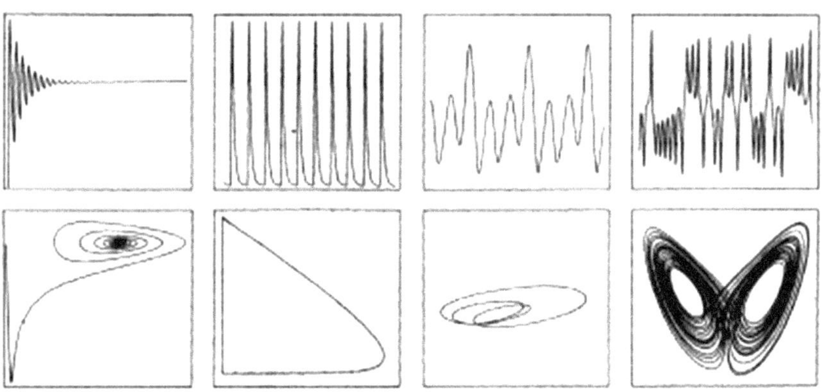

**Figure 4.7** ***Types of attractors.*** The top row shows traditional graphs indicating the signal amplitude over time for four different situations. The bottom row shows the corresponding phase diagrams, another way of graphing the information that reveals the type of attractor that governs the behavior of the system. (A) Point attractor, indicating a system that settles into a steady state in which there is no change. (B) Periodic attractor, indicating a system that repeats itself cyclically. (C) Complex periodic attractor, indicating a system that repeats itself periodically, here with a rhythm of 3. (D) Chaotic attractor that never repeats itself exactly but does show long-term trends. *(From Gleick J.* Chaos. *Viking; 1987. p. 50. Reprinted with permission of the author.)*

A more intricate type of complex system has a periodic or oscillating attractor and locks into a repeating cycle. Prey–predator relationships in animals frequently have periodic attractors. As the number of predators in a region increases (foxes, for example), the number of prey decreases (rabbits, for example), which in turn leads to a decrease in the number of predators, followed by a subsequent increase in the prey, and around and around the cycle.

The waves of the EEG in response to a conditioned stimulus also have a periodic-like attractor, caused by a cycle between excitatory neurons and the inhibitory ones that are present in most brain tissue.[13] Excitatory cells stimulate inhibitory cells, which in turn inhibit the excitatory cells, which then decrease the firing of the inhibitory cells, etc. By preventing runaway excitation, the inhibitory neurons help sustain focused neural activity.

A perturbation may alter the behavior of a system with a periodic attractor in the short run. However, as long as the perturbation is within certain limits, when it stops, the system will return to its habitual cycle. These limits define its *basin of attraction*, the set of initial conditions that will result in this behavior.

Even more unpredictable systems tend to be governed by *chaotic or strange attractors*. Contrary to its common use, the word *chaotic* in the context of complex dynamical systems refers to the presence of subtle long-range order that results from the complicated interaction of very large numbers of relatively sparsely interconnected parts or influences. Each neuron in the brain may connect with as many as 10,000 others. Yet because the density of neurons is so high in the brain, this nonetheless corresponds to sparse interconnection. (Systems that are too densely interconnected are more likely to fire as a unit, as occurs in epileptic seizures.) Over the short term, the behavior of a complex chaotic system appears random. It is necessary to follow its behavior over a long time to detect the order. The stock market provides a good example. Because of its sensitivity to so many different interrelated factors, it never repeats the same pattern. Nonetheless it shows various discernible cycles.

The complex interconnection of chaotic systems leaves them sensitive to small perturbations that may result in phase transitions. When one of its component levels undergos a phase transition, if conditions are right, it can drive the whole system to a less chaotic and more periodic state governed by a more predictable and familiar pattern. In contrast to noise, which is truly random, chaos can be turned off—that is, change into a *more* fixed pattern. This can be caused by even very small changes. These shifts are often predictable, which is what makes weather forecasting, for instance, possible but by the same token less than exact.

One way to think of a chaotic system is that it is balanced between, or on the cusp of, a number of different kinds of potential organization.[15]

Because of this readiness, systems with chaotic attractors are particularly sensitive to very small changes, which can alter the behavior of the whole system dramatically. Thus the popular notion that a butterfly flapping its wings in the Far East could cause a tornado in East Texas.

Freeman points out that resting brain activity is chaotic because it is composed of the neural activity of so many different sparsely interconnected components, each with many sublevels, and all with their own inherent neural rhythm.[13] Because they cannot settle on a common pattern of activity, in connection they share a chaotic rhythm. But instead of a handicap, chaos is what allows the brain to do its job.

Poised at rest on the edge of chaos, each subsystem of the brain as well as the brain as a whole can respond to known stimulation by reconfiguring its activity, or *self-organizing*, into a more stable, or more periodic-like pattern. The ability to flip from a resting chaotic state to any one of a large number of more ordered states—for instance, into the basin of attraction associated with dogs or into the basin of attraction associated with cats, or with rabbits—seems to be what allows the brain to perceive known distinctions and reproduce familiar patterns of behavior.

## RABBITS AND OLFACTORY PERCEPTION

The air is full of odors. The olfactory system must ignore most and instead focus only on those that carry important information.[13] Another challenge for the olfactory system is that each sniff excites only a small fraction of the receptors sensitive to any given odor.[21]

Yet *all* nasal receptors generate the same electrical response. The brain can recognize a given odor only from the neural activity initiated by the olfactory receptors it happens to stimulate. But if a different group of receptors is stimulated on each encounter, how does the brain know it is smelling the same smell? More technically stated, the question is, "How does the brain generalize over equivalent odors?"[13]

Freeman realized that another process, *odor perception*, is likely to intervene between olfactory stimulation and the behavioral response it evokes. The olfactory system is especially well-suited to study perception because only three synapses separate it from the emotional and memory centers of the limbic system (Chapter 5). Freeman and coworkers set out to understand the complex neural events that correspond to odor perception.[13,22] They followed both the EEG waves from a rabbit's olfactory system as well as the animal's behavior in response to different conditioned odors.

The rabbit's olfactory system consists of three main components, as shown in Fig. 4.8. The first layer consists of the nasal receptors, the odor-sensitive cells. The receptors synapse, or connect, with neurons of the *olfactory bulb* in the brain, the second layer. The neurons in the olfactory bulb in turn send axons to the *olfactory cortex,* the third layer.

The nasal receptors do not interact with each other, but nearby receptors tend to report to the same bulbar neuron. In contrast, neurons in both the bulb and the cortex interact with their neighbors. Each bulbar neuron sends information to a large number of cortical neurons. Feedback connections exist between the cortex and the bulb. In addition, both the bulb and the cortex are richly bidirectionally interconnected with the limbic system.

The researchers surgically implanted an array of 64 tiny electrodes in the olfactory bulb of rabbits. This allowed them to record the EEG activity in a

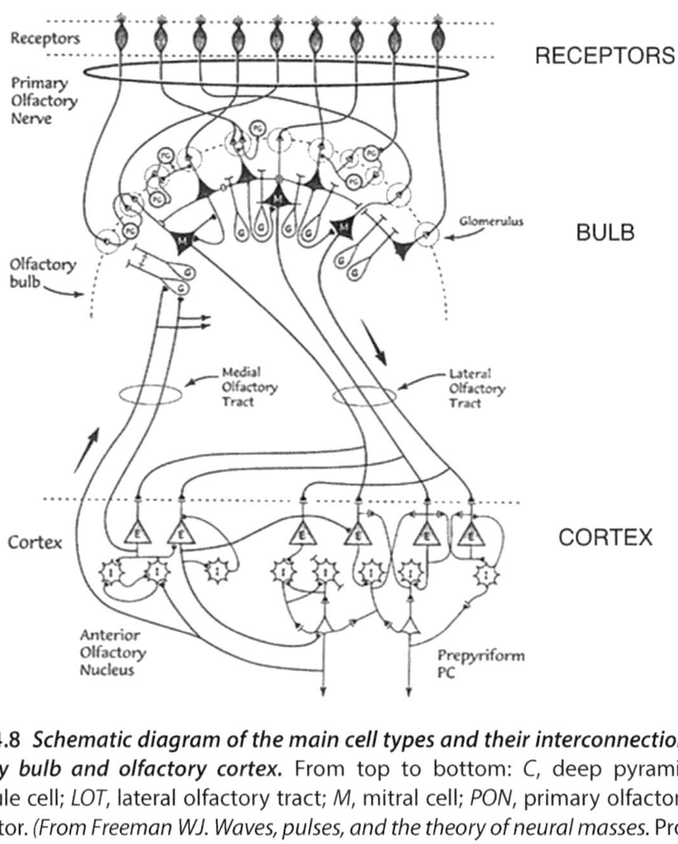


**Figure 4.8** ***Schematic diagram of the main cell types and their interconnections in the olfactory bulb and olfactory cortex.*** From top to bottom: *C*, deep pyramidal cell; *G*, granule cell; *LOT*, lateral olfactory tract; *M*, mitral cell; *PON*, primary olfactory nerve; *R*, receptor. *(From Freeman WJ. Waves, pulses, and the theory of neural masses.* Prog Theor Biol *1972;***2***. Reprint with permission of Elsevier.)*

sizable portion of the bulb in awake animals in response to a variety of different odors. Then they conditioned the animals to respond to several different smells, each with a different behavior: for example, to lick in response one odor and scratch in response to another.

They found that each sniff of a *conditioned* odor generated a series of fast or high-frequency gamma oscillations (30–80 Hz) in the EEG from the bulb that were more regular, or periodic, than its chaotic resting basal waves (Fig. 4.9, array on left). The pattern of the waves in the bulb was variable for different odors and even for repeated sniffs of the same odor. However, for

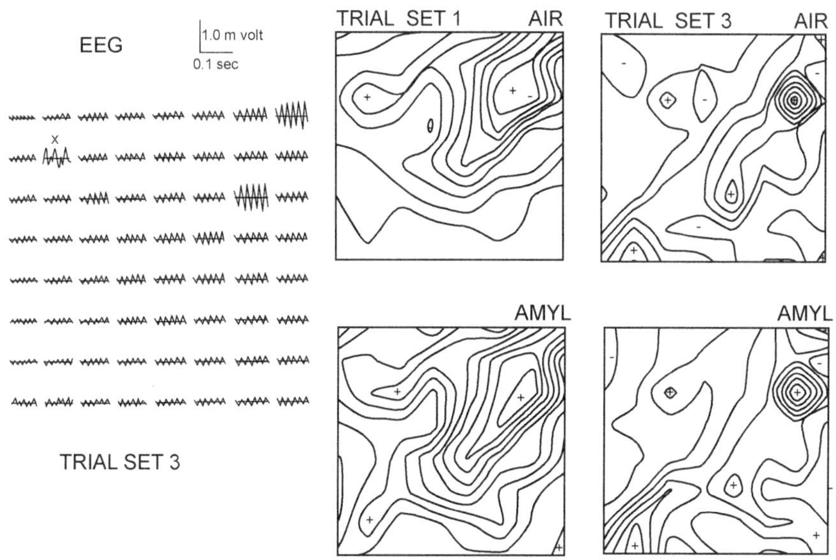


**Figure 4.9** ***Recordings from 64 electrodes on the surface of the olfactory bulb.*** (*Left*) During inhalation of an odor of interest to the rabbit, each of the 64 electrodes on its olfactory bulb shows short bursts of oscillation in the gamma range of the electroencephalogram. They all demonstrate more or less the same waveform, reflecting a response of the bulb as a whole. What differs between the different electrodes is the amplitude of the gamma waves. (*Right*) When the 64 amplitudes are plotted together as contour plots showing the peaks and valleys of amplitude in response, they reveal spatial amplitude modulation pattern (AM patterns) or how the amplitude of the waves vary from place to place on the bulb. The AM pattern to the background, air (AIR), and to the stimulus, amyl acetate (AMYL) are different from trial set 1 to trial set 3, even though the stimulus was the same. The pattern for the background also changes. The changes indicate that brain activity, which carries the meaning of the stimulus to the animal, is altered by learning. *(From Freeman WJ, Schneider W. Changes in special patterns of rabbit olfactory EEG with conditioning to odors.* Psychopysiology *1982;**19**(1):44–56. Reprinted with permission from Elsevier.)*

any given sniff, each of the 64 electrodes always reported roughly the same waveform. Even though only a small proportion of the receptors for the odor was stimulated, and a different group each time, *all* neurons in the bulb responded in the same way on each sniff. This signifies phase transition, the self-reinforcing response of the layer as a whole.

Information about the spatial array of *nasal receptors* that fired, which had to carry the identity of the odor, could not depend on which bulbar neurons fired because they all did to a greater or lesser extent. Instead, what differed from electrode to electrode was the amplitude, or the height, of the waves, recorded by the different electrodes. Each odor was characterized by a specific pattern of EEG amplitudes over the 64 electrodes, or over the bulb as a whole. The various amplitude patterns can be plotted like a topographic map of a mountain range. A few of the patterns are shown in Fig. 4.9 on the right.

Generalized arousal via norepinephrine, along with the release of other neuromodulators molecules such as dopamine (which signals the expectation of reward or punishment) is required for a phase transition. By augmenting synaptic activity in response to the odor, neuromodulators increase the gain, or the number of action potentials a neuron receives from neighboring neurons for each of its own pulses. The large extracellular current that results enhances the possibility of a bulb-wide response to an odor.

After only a few sniffs of a conditioned odor, the intensity of the summed currents in the extracellular fluid can make neurons in the bulb start responding to signals from each other rather than from the other smells in the environment. The researchers found that the response of the entire bulb to an odor occurs when as few as 1% to 5% of its neurons are connected to each other in cell assemblies. This phase transition drives the whole olfactory system to a more periodic, or stable, state than before, which is an important step in perception.

This does not mean that all the bulbar cells fire at the same time. Rather they begin to function as a coordinated population. Freeman stresses the distinction between the *microscopic level*, in which neurons act individually, from the *mesoscopic level*, in which the same neurons act as part of a population. Although each neuron largely maintains autonomy, it also yields a bit of its individuality to the brain to do the work of the collective.

Other unreinforced odors still cause specific neurons in the bulb to fire. However the information is drowned out by the coordinated response of the bulb as a whole to the reinforced odor. Freeman emphasizes that sensory input is a local property that records the features of the environment; it slowly modifies neural connections over time. Perception, in contrast, is a global process that makes a *meaningful* connection between the inner and outer world of an organism based on its needs and prior experience. It depends on the organization of neurons into coordinated populations. It also depends on the ability of these populations to broadcast their output to each other through their many reciprocal connections.

The researchers found that over a few weeks, the neural pattern in the bulb drifted somewhat for any specific conditioned odor. Although the animal's neural response to the odor had changed, its behavioral response remained constant. Somewhat larger changes in the pattern occurred when the reinforcement associated with an odor was changed. Even more dramatic changes occurred when an animal was conditioned to a new smell. The pattern of activity of its bulb to all other odors would also change (Fig. 4.10). However, again, the animal's behavior in response to each of the known smells stayed the same. Even the neural pattern for the

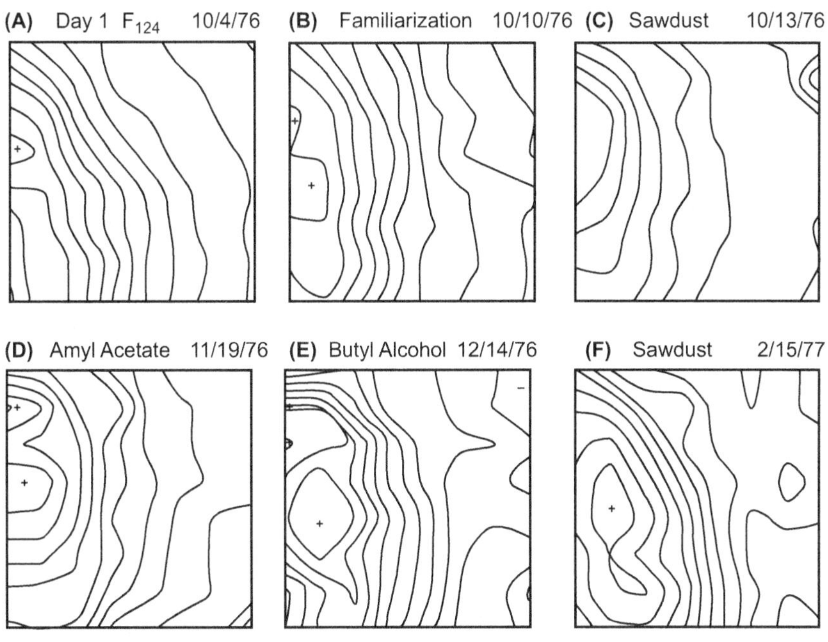


**Figure 4.10** ***All the amplitude modulation (AM) patterns change whenever the rabbit is conditioned to respond to a new smell.*** First the animal became familiar with the experimental setup (A, B). They were then conditioned to sawdust (C), about a month later to amyl acetate (D), and a month after than that to butyl acetate (E). When the animal was trained and tested 2 months later with sawdust, its AM pattern was different from the initial AM pattern to sawdust (F). *(From Freeman WJ, Schneider W. Changes in special patterns of rabbit olfactory EEG with conditioning to odors.* Psychopysiology *1982;**19**(1):44–56. Reprinted with permission from Elsevier.)*

background (the response to air alone), a signature that differed for each rabbit, drifted over time.

Freeman points out that the drift in the patterns over time, as well as the more dramatic changes that occur with newly conditioned smells, reflect the continuous modification of the neurophil, the allover synaptic connections, due to the intervening experience of the animal. This cumulative synaptic record is required for the animal to recognize any specific odor over time. Even broader changes occur when the directly relevant set of experiences has expanded.

Yet if the rabbit had only the neural activity pattern to go on and this pattern had changed over time, how did it *know* it was the same odor? Freeman's work shows that the initial EEG pattern of amplitudes in the bulb and all the subsequent ones for a given conditioned smell all fall within the same chaotic attractor in the olfactory cortex.

Transmission of the consensus signal from the olfactory bulb to the olfactory cortex occurs via action potentials of individual cells. Each bulbar neuron sends information through its axon to many widely separated cortical neurons. Every cortical neuron in turn sums the information it receives from many different bulbar neurons. This *divergent–convergent* connectivity further amplifies the consensus message carried by the common gamma waves from the bulb. The large extracellular current generated by the consensus message causes a phase transition of the olfactory cortex, driving it into the basin of its attraction for that smell. The cortex also has an attractor for each conditioned smell, which together form an attractor landscape.

In contrast, neural firings related to the unreinforced odors tend to cancel each other, further reducing the immediate impact of the various unreinforced signals. If circumstances change abruptly and the ignored information becomes important, Freeman argues that it can be transmitted by phase transition to the cortex. For instance the whiff of food that goes unperceived in a hungry animal because of the prior whiff of a predator may become important once the predator has safely passed.

The sensitivity of the brain's resting chaotic rhythm to small changes also allows it to undergo phase transition and to self-organize in response to potentially important *novel* input. In addition, it introduces some unpredictability and disorder, which is critical to the novelty often required for successful trial and error behavior and also sometimes has a role in intuition. Chaos provides the brain with the balance it needs between stability and flexibility.

If a new odor is introduced to a rabbit and it evokes a slight reinforcement, either positive or negative, it will move the olfactory bulb into a chaotic state slightly different from the background one. More reinforcement will cause enough cells to join together to ignite the entire bulbar layer and drive the whole olfactory system into a new more periodic-like pattern governed by a new attractor. Repeated encounters with the odor deepen the basin of attraction, making it easier for the animal to discriminate the smell.

The animal's cell assemblies come to reflect the cumulative history of its encounter with an odor over time. Any encounter with it will send the cortex into the attractor for that smell. Because of current needs signaled by the value system and past experience encoded in synaptic memory, a relatively weak smell might trigger a response of the bulb as a whole and be perceived, whereas a stronger one might not. The emotional/motivational system confers *intentionality*, which according to Freeman signifies that perception is shaped by meaning as it relates to an organism's goals and its experience.

In the wild, drives such as hunger generate the arousal and the intention that directs the animal's behavior toward satisfying its needs, such as going to the place where it is likely to find food. The expectation of food in turn presensitizes it to the smells it will likely encounter, probably by enlarging or deepening the basins of attraction.[13] In the laboratory, however, the rabbits may need a mild shock to ensure their arousal, and thus their interest, before they are presented with the test odors.

Freeman has shown that each of the other sensory cortices can also undergo phase transitions by sensory information that holds meaning for an animal.[23,24] The sensory cortices in turn send the reinforced sensory information to other areas of the brain. Each has its own attractor landscapes. Together they encode a multisensory perception that is situated in time and space and is rich with information about the successes of past actions and the probable consequences of future action.

Freeman proposes that each cerebral hemisphere also forms its own attractor landscape.[24] Neural activity carried by the sensory-driven amplitude patterns condenses into self-organized, hemisphere-wide configurations of activity, or what he calls *wave packets*, that oscillate in the gamma and beta wave range. (Beta waves go from 14 to 26 Hz and are generally associated with concentration, which need not be conscious.) They come in frames that last about a tenth of a second and then break up, ensuring that perception always remains fresh.

Consciousness also corresponds to an overall attractor landscape according to Freeman. Like each successively higher-level in the brain, consciousness in some sense "enslaves" the lower ones because of the complex interconnectivity between higher and lower anatomical regions, as well as higher and lower levels of functional organization. Consciousness, like each higher level, helps order as well as contextualize the functioning of subordinate levels.

Rabbits may well have primary consciousness, or consciousness of the current moment. However they are unlikely to cognize at a conscious level the fertile meaning implicit in their percepts. Yet Freeman's model gives even the lowly rabbit the ability to form ever-varying sensory-based megapatterns that integrate their experience of a specific odor.

Freeman and his colleagues also demonstrated the appearance of more abstract perception—perhaps even insight—in gerbils, although again this perception is unlikely conscious. When gerbils were reinforced for recognizing specific ascending from specific descending auditory chords, eventually they spontaneously came to distinguish whether novel chords were ascending or descending.[25] This indicates they developed separate attractors for ascending and for descending cords. A much more advanced and subtle version of the same ability for abstraction likely has a role in allowing a scientist to intuit the megapatterns that can underlie widely variant phenomena.

Demonstrating insight or intuition in humans by these techniques lies in the future. However Freeman extended his sensory studies to humans and also explored more integrated levels of cognitive function in human brain activity.[26] Even in humans, he argues, the perception indicated by the coherent gamma activity he records is *not conscious* perception. Rather it occurs before awareness and corresponds to unconscious activation and unconscious recognition of the fullness of potential meaning, although portions of this meaning may be experienced intermittently at a conscious level. Likewise, even in humans this perception is directed by an intention that is more basic and complete than what we experience as conscious will.

Freeman's work was ahead of its time at many different levels. To summarize the current status of only a few: Researchers increasingly accept the importance of chaos in the brain.[27,28] Moreover the brain's internal energy is now well studied.[17] Researchers are also becoming more interested in response variability as it relates to more intrinsic aspects of brain activity.[29] However many workers still think that synchronous gamma oscillations the brain in general correspond to either feature binding or to consciousness[30] rather than the perception, or meaning making, below awareness that Freeman champions. Freeman, in contrast, argued that consciousness corresponds to widespread phase synchrony in all EEG bands.[26]

Workers who study olfaction are now open to the idea that the bulb has an active role in odor recognition, as Freeman proposed.[31] Interestingly, recent studies have shown that olfactory discrimination is not completely dependent on gamma waves, at least in mice; it can still occur when they are abolished.[32] Yet it is much slower and cruder. This highlights the importance of phase transition and attractor capture for efficient olfactory discrimination and for the rich meaning-making that Freeman's work describes.

## INTUITION AND THE SELF-ORGANIZING BRAIN

Self-organizing systems reflect a confluence of both bottom-up and top-down forces. With a *mechanical* treadmill, running faster makes the treadmill go faster, which in a circular fashion in turn supports running faster. Likewise, the feedback resulting from the integrated function of the brain turns bottom-up processes into top-down, and vice versa.[33] As a consequence, the distinction between top-down and bottom-up influences in some sense disappears.

The reciprocal connections among neurons in a layer, among individual tissue layers, and among various regions allow the brain to function as a well-integrated and extremely discriminating whole. The process of meaning making is governed by local synaptic connections at many different interactive levels in reciprocal interaction with the emotional/motivational system and its neuromodulators, which guide intentionality. What this system considers to be of value strongly influences synaptic connectivity, which in turn determines what the sensory systems can respond to and thus what is available for the emotional/motivational system to achieve its goals. Both ordinary perception and intuition depend on this complex *circular* way the brain has of making meaning.

Freeman argues that because of this circularity, the intentionality that guides meaning making is more *unified* than conscious volition or intention.[13] It automatically brings the body and all of its senses together with the mind in order to anticipate and pursue the organism's needs and goals. It is also more *whole* than conscious intent in the sense of automatically bringing all of the organism's knowledge and experience to bear on its purposes. Freeman's research, which highlights the brain as a self-organizing meaning-making system, argues for the intuitive basis of all experience.

I propose that the biologically tethered *sphere* of mind his work evokes, which shapes perception below the level of consciousness based on meaning, is the basis of an *intuitive or implicit self*. A high-level, self-organized activity that precedes awareness and is responsible for our intentionality, the intuitive self hands the conscious mind the world it perceives. It is in touch with a ground of experience that is much fuller than the conscious mind can possibly capture. During moments of intuition, we sometimes get to see more of this fullness than we usually do. You might get a partial glimpse of the importance and prowess of the intuitive self by looking around a room you know well *very slowly* and asking: What is it that makes your eyes focus on the details it does, rather than on others?

The intuitive or implicit self is as central to structuring higher-level cognitive meaning as it is to the more elementary meaning making that guides conscious attention as one looks around the room. Sophisticated meaning making, like lower-level meaning making, is grounded in perception that is largely given to the conscious mind, based on needs, goals, and the totality of a person's past experience or knowledge. The novel and highly significant perception that characterizes intuitions, like all other perception, depends on intricate feedback between top-down and bottom-up levels. The intentionality behind an intuition drives what strands are included, but, in a chicken and egg fashion, the potential meaning of the information or knowledge to be included drives the intuition. As in all self-organizing systems, top-down and bottom-up influences converge.

The ability of the brain to reorganize into a novel state in response to small changes, a hallmark of chaotic self-organizing systems, accounts for many examples of intuition. A brief encounter, such as seeing a baby placed in its bath, or the memory of something seen or read weeks ago, might now cause a significant reordering of something apparently unrelated and previously ignored but newly important. Of course the intuition may well depend on many more subtle changes in the cognitive landscape that occurred in the interim as a result of apparently unrelated experience.

It is possible to get a hint of the chaotic sensitivity of the brain to small changes when a very slight or even intangible difference in the way we hear something makes us realize, without knowing why, that what we had assumed meant one thing actually means something very different we had not considered before. These reconfigurations are likely to occur only if they are consistent with our underlying needs or goals. Bottom-up and top-down forces must converge. Freeman's work suggests that meaning, which makes up minds, emerges from the complex circular causality of the interconnected brain.

The mind is sometimes forced to respond to large changes in internal or external conditions. Large-scale reorganization of the neural and cognitive landscape is often required when past meaning making proves seriously inadequate to current challenges. After a deep struggle to make sense of a matter, whether it is personal or intellectual, we sometimes find our thinking radically rearranged. During intuition at its richest, level upon level of understanding falls into place. For an instant at least, we have access to the whole of the issue, as well as its connection to many other related issues. In these moments, much more of the meaning-making substratum of mind than usual is available to awareness. This radical reorganization can take time, and in the interim it may cause a certain amount of cognitive discomfort.

Especially penetrating intuitions can reorganize the activity of many different aspects of mental function. They work together to restructure a large portion of the cognitive landscape into a much more coherent global pattern than the conscious mind can often appreciate in the moment. Understanding the brain/mind as a global self-organizing system structured by a more integrated meaning-making force than the conscious mind can begin to explain one of the richest as well as the most perplexing aspects of intuition: how can it sometimes combine so many apparently different aspects of experience that often seem opposed to each other in conscious thought? For instance, the mental, the bodily, the metaphoric, process, and content all came together as I looked inside in response to my friend's daunting question, What is intuition? Encounters such as this provide evidence for the underlying wholeness of the intuitive meaning-making mind.

## REFERENCES

1. Rosch E. Natural categories. *Cogn Psychol* 1973;**4**:328–50.
2. Murphy G, Lassaline ME. Hierarchical structure in concepts and the basic level of categorization. In: Lamberts K, Shanks D, editors. *Knowledge, concepts and categorization.* Cambridge (MA): MIT Press; 1997. p. 93–131.
3. Rosch E, Mervis CB, Gray WD, Johnson DM, Boyes-Braem P. Basic objects in natural categories. *Cogn Psychol* 1976;**8**:382–439.
4. Bechtel W, Abrahamson A. *Connectionism and the mind.* Hokoken (NJ): Wiley-Blackwell; 1991.
5. Smolensky P. On the proper treatment of connectionism. In: An A, Goldman AI, editors. *Readings in philosophy and cognitive science.* Cambridge (MA): MIT Press; 1993. p. 769–99.

6. Churchland P. *A neurocomputational perspective: the nature of mind and the structure of science.* Cambridge (MA): MIT Press; 1992.
7. Goreman RP, Sejnowski TJ. Learned classification of sonar targets using a massively parallel network neural networks. *IEEE Trans Acoust* 1988;**36**:I135–40.
8. Contrell GW. Extracting features from faces using compression networks: face, identity, emotion, and gender recognition using holons. In: Touretzky DS, Elman JL, Sejnowski TJ, editors. *Connectionist models: Proceedings of the 1990 summer School.* San Mateo (CA): Morgan Kauffmann; 1990.
9. Elman JL. Finding structure in time. *Cogn Sci* 1990;**14**:179–211.
10. Cleeremans A, McClelland JL. Learning the structure of event sequences. *J Exp Psychol Gen* 1991;**120**:235–53.
11. For a paper describing how neural networks are being used to decode signals sent by the brain see Miller K.J., Schalk G, Fetz EE, Nijs D, Ojemann, Rao RP. Cortical activity during motor movement, motor imagery, and imagery-based online feedback. *Proc Natl Acad Sci USA* 2010;**107**:4430–5.
12. Macknik SI, Martinez-Conde SM. The role of feedback in visual attention and awareness. In: Gazzaniga MS, editor. *The new cognitive Neurosciences.* 4th ed. Cambridge (MA): MIT; 2009. p. 1163–75.
13. Freeman WJ. *How brains make up their minds.* London: Weiden & Nicolson; 1999.
14. Hubel DH, Wiesel TN. *Brain and visual perception: the story of a 25-year collaboration.* New York: New York University Press; 2005.
15. Freeman WJ. Chaos in the CNS: theory and practice. In: Greenspan RJ, Kyriacou CP, editors. *Flexibility and constraint in behavioral systems.* New York: John Wiley; 1994. p. 177–84.
16. Wilf M, Strappini F, Golan T, Hahamy A, Harel M, Malach R. Spontaneously emerging patterns in human visual cortex reflect responses to naturalistic sensory stimuli. *Cereb Cortex* 2017;**27**(1):750–63.
17. Raichle ME. The brain's dark energy. *Sci Am* 2010;**302**:44–9.
18. Freeman WJ, Skarda CA. Mind/brain science: neuroscience on the philosophy of mind. In: Lepore E, Gulick R, editors. *John Searle and his critics.* Oxford (UK): Blackwell; 1990. p. 121.
19. Heon-Jin L, Macbeth AH, Pagani J, Young WS. Oxytocin: the great facilitator of life. *Prog Neurobiol* 2009;**88**(2):127–51.
20. Strogatz SH, Stewart I. Coupled oscillators and biological synchronization. *Sci Am* 1993;**269**:102–9.
21. Complicating the situation even more, evidence suggests that most odorants bind to more than one type of olfactory receptor, and individual olfactory receptors bind with varying affinities a number of different odorants with partial chemical similarity. Araneda RC, Peterlin Z, Zhang X, Chesler A, Firestein S. A pharmacological profile of the aldehyde receptor repertoire in rat olfactory epithelium. *J Physiol* 2004;**555**(3):743–56.
22. For a seminal article on this work see Freeman WJ, Skarda CA. How brains make chaos in order to make sense of the world. *Behav Brain Sci* 1987;**10**:161–95.
23. Ohl F, Deliano M, Scheich H, Freeman W. Early and late patterns of stimulus-related activity in auditory cortex of trained animals. *Biol Cybern* 2003;**88**(5):374–9.
24. Freeman WJ. The wave packet: an action potential for the 21st century. *J Integr Neurosci* 2003;**2**:3–0.
25. Ohl FW, Scheich HW, Freeman WJ. Changes in patterns of ongoing cortical activity with auditory category learning. *Nature* 2001;**412**:733–6.
26. Pocket P, Bold GE, Freeman WJ. EEG synchrony during a perceptual-cognitive task: widespread synchrony at all frequencies. *Clin Neurophysiol* 2009;**120**:695–708.
27. Kitzbichler M, Smith M, Christensen S, Bullmore E. Broadband criticality of human brain network synchronization. *PLoS Comput Bio* March 20, 2009;**5**(3):e1000314.

28. Papo D, Goñi J, Buldú JM. On the relationship of dynamics and structure in brain networks. *Chaos* 2017;**27**(047201). https://doi.org/10.1063/1.4981391.
29. Fox MD, Snyder AZ, Zacks JM, et al. Coherent spontaneous activity accounts for trial-to-trial variability in human-evoked brain potentials. *Nat Neurosci* 2006;**9**:23–5.
30. Singer W, Gray CJ. Visual feature integration and the temporal correlation hypothesis. *Annu Rev Neurosci* 1995;**18**:555–86.
31. Devere S, Rinberg D. Dark matter of the bulb. *Nat Neurosci* 2014;**17**(4):485–7.
32. Abraham MN, Egger V, Shimshek DR, et al. Synaptic inhibition in the olfactory bulb accelerates odor discrimination in mice. *Neuron* 2010;**65**:399–411.
33. Goerner SJ. Chaos, and deep ecology. In: Abrahams FD, Gilgen AR, editors. *Chaos theory and psychology*. Westport: Greenwood/Praeger; 1995. p. 17–38.

# CHAPTER 5

# Emotion and Motivation

## Contents



We bare our teeth in anger just as dogs do. Despite our capacity for lofty feeling, as Darwin[1] recognized, we live and die as animals. Even our cherished cognitive capacity is to a large extent a derivative of our emotions. In turn, emotions depend on a series of pathways that evolved in much simpler species and greatly increased the probability that individual organisms as well as their species would survive. The success of these pathways ensured their retention within evolution and made their further elaboration likely.

Emotions supply the basic impetus for all behavior and thought. They make things matter. Because they determine what is important, or what has value, they underlie and guide all intuition, be it personal, intellectual, or artistic.

What has meaning or value includes not only what is recognized as such at a conscious level. For humans, just as for other animals, it also includes meanings and values fostered below awareness by our biological organism as it interacts with our experience. These hidden emotional meanings and valences often have a pivotal role in intuition.

## EMOTIONS AND EVOLUTION

Consider the emotional life of frogs. They perceive anything that is small and moves in stop-and-start fashion, such as an insect, to be food and attempt to capture it with their tongue. To avoid danger, they always jump

*Understanding Intuition*
ISBN 978-0-12-814108-3
https://doi.org/10.1016/B978-0-12-814108-3.00005-1

to where it is darker. Likewise, their sex life depends on simple auditory and tactile cues.[2]

What humans experience as emotion began in evolution as instinctual patterns of bodily arousal triggered by specific environmental clues. These primitive emotions, or drives, mobilized various bodily systems in a coordinated way that allowed organisms to be more successful at obtaining food, reproducing, attacking a foe, gathering information, and/or getting out of a predator's way. Simple emotional responses were in place well before reptiles appeared. However, a significant portion of the evolving brain circuitry became devoted to these powerful adaptations when reptiles were on evolution's cutting edge.[3]

The brain stem, which is pivotal to reptilian functioning, governs the bodily arousal that is at the core of emotion, as well as many other basic aspects of involuntary functioning. It works in conjunction with the nearby hypothalamus, a major control station for the autonomic nervous system. Together they coordinate the automatic adjustments the body makes in response to significant changes in the internal and external environment.

Emotional pathways became more elaborate during mammalian evolution. Mammals developed increasingly complex sensory and cognitive mechanisms. Connections were forged between these newly evolving brain structures, which allowed animals to discriminate, record, and recall their experience more successfully, and the older brain stem structures responsible for the bodily changes associated with emotions. Thus the sensory mechanisms necessary for increasing complex discriminations as well as the brain structures necessary to remember the information all coevolved with the emotional pathways. Many of the newly evolved cognitive structures are found in the limbic system, which is often seen as the center of emotional activity.

Mammals retained many of the older automatic physiological arousal mechanisms. Yet the specific environmental cues that triggered arousal, as well as the specific behaviors they evoked, came to differ widely. These depended on the habitat, predators, and prey of the species, as well as the experience of individual animals. Mammals could increasingly modulate a portion of their behavior in the light of past experience. Emotions and drives such as fear and hunger were now no longer necessarily locked into invariant patterns.

An animal might learn to reduce close encounters with danger by associating fear with the sound and smell of a predator. When strong fear was triggered, its brain also would automatically remember as well as attach emotional valence to the context in which it occurred. This helped the animal avoid similar circumstances in the future. For example, these *conditioned* as opposed to innate emotional signals might make it learn to favor the particular approach to the river where it was less likely to encounter the smell of its predators.

With human evolution, the emotional pathways became integrated with the growing neocortex, especially the burgeoning prefrontal region. Compared with the brain of chimpanzees, our closest kin, the human brain differs primarily by its expanded prefrontal region, along with adaptations for speech. The prefrontal region governs working memory, directed attention, as well as the capacity to plan, evaluate, execute, and inhibit behavior. The evolving capacity to plan conferred by the enlarged prefrontal region allowed early humans to become increasingly emancipated from the struggle to satisfy momentary needs. However the close connection between the older emotional centers of the brain and the higher cognitive regions ensured that emotions would continue to bias mental activity in line with survival needs.

Many of the basic physiological events triggered by a primary, or innate, emotional response have been passed down relatively intact from lower animals to humans. With aggression, our brain and body receive a pulse of adrenaline as well as other hormones that provide a substantial rush of energy. With fear, we are subject to a different cocktail of hormones that may cause us to flee or, alternatively, to heighten alertness and fixate on the source of danger. Like a deer caught in headlights or a squirrel sensing it is observed, we may freeze for a moment to become less noticeable and collect information.

## HUMAN EMOTIONAL CUES

Just like other animals, humans have an innate tendency to respond to certain environmental cues with a strong emotional response. These primary *inducers* proved either particularly advantageous or detrimental to survival during evolution. Spiders and snakes are some examples of stimuli we are apparently *genetically prepared* to fear.

Also like other animals, we are born with a number of genetically programmed drives (strong seeking or avoidance urges) that ensure our instinctual needs are satisfied.[4] For example, newborns seek and respond with pleasure to being held and other forms of gentle physical contact.

The infant's need to be held is reinforced by the connection between feeding and holding, but it also appears to be a basic drive in itself.[5] Babies who are adequately fed but seldom held tend to have unusually strong stress reactions and are predisposed to mental illness as they mature.[6] The absence of physical contact can also cause structural alterations in the developing brain.[7] The need for physical contact shades into the need for emotional connection and comfort. Together they motivate much of human behavior throughout life.

Some emotion researchers have posited that various *patterns of stimuli* act as primary inducers of different emotions.[8] Fear is triggered when too much information arrives in too short a time (such as a looming figure or a loud noise) and pleasure, when the flow of novel information is just right. Contentment, in contrast, occurs when there is a decrease in the level of stimulation or stress.[9] Emotion researcher and physician Antonio Damasio[10] captured all of this and more by framing the emotional or affective system in terms of bodily homeostasis. He proposed that humans, like other animals, respond with emotion to inner or outer events that strongly influence physiological well-being.

Human affective responses become associated through conditioning with a much broader range of stimuli. Certain categories of events become linked with pleasurable bodily states, and others with unpleasant ones. Over time we each develop a highly personalized profile to our emotional needs and responses. Even objects or situations that appear neutral to the conscious mind may through association become tinted below awareness with a sense of bodily preference or aversion, of "goodness" or "badness." Even objects or events that are only imagined in the mind come to generate emotional feelings. On an especially problem-fraught and frantic afternoon, the reverie of lying on the sand in the summer sun might bring a conscious sense of pleasure or lighten a bleak mood.

Likewise, artists employ images, both realistic and imaginary, to evoke complex mixtures of conscious and unconscious emotions, Much of the power of art comes from its ability to induce subtle layers of feeling. Poetic imagery, for example, very often evokes the web of emotion-tinged meanings that give experience its force.

## THE PRIMARY FEAR PATHWAY

One evening I found myself suddenly in the midst of a full-blown fear reaction with no notion of why. Eventually I caught the subliminal trace of something small and fast with many legs running across the surface of my dresser. (Yet I thought I was not afraid of spiders, because I often peacefully coexist with many!)

This intense response indicates direct activation of the amygdala, the structure in the brain that initiates the fear response. The thalamus, an early sensory relay station in the brain, sends all incoming sensory information separately to both the amygdala and the various sensory areas of the cortex. When the amygdala detects a primary inducer of emotion, it triggers a primary fear response and generates full bodily arousal before higher-brain centers have a chance to determine the specific nature of the threat.

The amygdala has limited discriminatory capacity. It is genetically prepared to recognize not objects, but rather certain features of objects. For example, it reacts to snake-like motion but not necessarily to sleeping snakes, and apparently it reacts to scurrying spiders, not sleeping or spinning ones. The amygdala has some memory capacity and can generate a fear response through conditioning to objects or events that become strongly associated with a primary fear-inducer.[11]

A small almond-shaped structure, the amygdala is attached to the hippocampus, which is critical to the formation of memories (Fig. 5.1, left). In addition, it is richly connected to the hypothalamus and brain stem, both of which control automatic bodily responses (Fig. 5.1, left). Although it also functions in other emotions, its role in the fear response is best understood.

The amygdala generates three broad kinds of responses when activated by a fear-laden cue[8]: (1) It alters the internal milieu and the state of the internal organs by mobilizing the autonomic nervous system and its connections to the hypothalamus and the brain stem. For example, it increases the heart rate and diverts blood flow to the large muscles. (2) It generates fear-related behaviors by stimulating the release of blood-borne molecules such as stress hormones and other neuromodulators in the brain and body. (3) It enhances mental alertness, and thus the ability to take in and integrate information about the source of fear and decide a course of action, via the release of the nonspecific neuromodulators that influence synaptic activity in large swatches of the brain. The three kinds of responses set in motion by the amygdala (changes in the internal milieu and internal organs, changes in behavior, and changes in cognition) all feedback on and amplify each other.

An emotion is experienced as a *conscious feeling* when the internal and external bodily changes associated with it register in the various body-sensing regions in the brain and are subsequently represented at a conscious

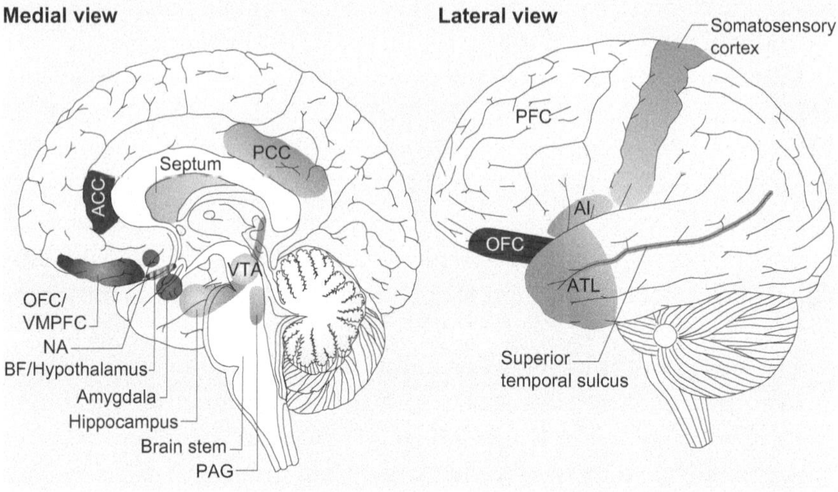


**Figure 5.1** ***Core and extended regions of the human emotional system.*** Some regions can be considered core emotional areas (light brown). At the subcortical level, they include the amygdala, the nuclear accumbens (NA), and the hypothalamus. At the cortical level, they include the orbital frontal cortex (OFC), the anterior cingulate cortex (ACC), and the ventromedial prefrontal cortex (VMPFC). Others can be considered part of the extended emotional system (light gray). At the subcortical level, these regions include the brain stem, the ventral tegmental area (VTA) and the associated mesolimbic dopamine system, the hippocampus, the periaqueductal gray (PAG), the septum, and the basal forebrain (BF). At the cortical level, they include the anterior insula (AI), which is noted on the surface but is hidden within the lateral sulcus close to the secondary somatosensory area, the prefrontal cortex (PFC), the anterior temporal lobe (ATL), the posterior cingulate cortex (PCC), the superior temporal sulcus, and the somatosensory cortex. *Adapted from Pessoa L. On the relationship between emotion and cognition. Nat Rev Neurosci 2008;9:148–58, with permission from Nature Publishing Group.*

level. Somatic information from a variety of different kinds of receptors is supplied to multiple bodily integration sites, or *maps*, in the nervous system from the spinal cord up to the neocortex. Together they provide a moment-to-moment online picture of the internal and external state of the body. Feeling is the brain's version of the body.[8] With the primary emotional response, conscious awareness of arousal and other body changes may occur only after more immediate priorities have automatically been taken care of. As William James[12] stressed, first we run from the bear and then we know we are afraid.

Together, three of the higher-level somatic integration sites are known as the somatosensory cortices[8] (Fig. 5.1, right). One is the primary somatosensory strip, or *S1*, which is on the outside of the brain adjacent to the primary motor strip. It registers information from the muscles as well as from the surface of the body, such as touch and temperature. It works in close conjunction with the secondary somatosensory area, called *S2*.

The insula, the third site, is hidden along with S2 within the lateral sulcus, or grove, of the brain (Fig. 5.1, right). Its posterior portion is the primary area in the cortex for registering most internal body sensations. The anterior portion provides an interface for bodily sensations and cognitive experience. The insula is important in many feeling states associated with emotion.[13] It is also part of a *salience network*, which determines what we attend to and whether our attention is directed internally or externally.[14] The insula has an important role in many aspects of intuition.[15]

## THE SECONDARY FEAR PATHWAY

The primary emotional pathway is augmented by the secondary emotional pathway.[8] Its complexity is best conveyed by an example. A woman's first important romantic relationship broke up the night she and her boyfriend saw a *Star Trek* movie. Many years later, she happens to see a snippet of the movie on a TV in the electronics section of a big-box store. As she begins to tell her current partner her opinion of the movie, her thoughts flash back to that difficult evening. These memories fade into the background as she begins to remember the plot of the movie. Nonetheless her breathing becomes constricted, she starts to sweat, and her mouth goes dry.

In the secondary emotional pathway, environmental information generates an emotional response only after it is processed in the cortex and its emotional meaning has been assessed. The prefrontal cortex juxtaposes the current experience with memories associated with similar kinds of experiences from the past, along with the emotion associated with them. In the situation just described, memories about other breakups would be activated even if they did not come to awareness.

After this assessment, the prefrontal cortex together with the amygdala initiate the fear-related arousal along with the other bodily changes and the heightened mental activity associated with fear. The bodily changes are in turn relayed through both nervous and hormonal pathways to the somatosensory regions of the brain. Again, these alterations in body states represented by the brain are largely what *is felt* at a conscious level in a secondary emotional response.

The emotional experience can in turn further modify what comes to mind. The woman in the big-box store might begin to have vulnerable, angry, or even vindictive thoughts, and each would in turn influence the body in somewhat different ways. A resonating loop develops between the body and brain, which may alter even as it sustains and perhaps amplifies the emotional response. Although the secondary fear reponse generally remain weaker than the primary fear response, it may be substantial.

Thoughts or memories about the *current* object or situation need not act as the emotional trigger in the secondary emotional pathway. More distantly associated emotional memories might initiate the emotional response. A sequel to the movie seen the night of the breakup, even though most of the cast was different, might have set the process in motion.

Importantly, the secondary emotional response can occur covertly. It does not require conscious awareness of the trigger, of the emotion-laden memories associated with it, or even that an emotional response is occurring. The woman might have begun to experience the physiological symptoms even if she was unaware that the snippet she caught on TV was from that particular movie, she did not remember that she had seen the movie with her old boyfriend, and if the memories of the breakup that were activated had not come to awareness.

Moreover implicit memories, or memories that have never been available to awareness, can lead to a secondary emotional response, as the work of Joseph LeDoux first showed.[11] It can also occur in response to subliminal stimuli. In addition, a delay might intervene between the environmental trigger and the reaction. Not infrequently it is impossible to pinpoint what is causing a secondary emotional reaction.

Secondary emotional responses that are low-level can nonetheless influence subsequent behavior. One study found that *subliminal* images of naked people could attract or repel the subsequent placement of attention of subjects depending on their erotic preference.[16]

Even enthusiastic champions of the bodily pathway accept the idea that sometimes the bodily reaction with secondary emotion might be only apparent—that is, the experience does not cause a full body response. In what Damasio[8,10] called the "as-if pathway," a frequently encountered stimulus that triggers the secondary emotional pathway evokes the experience of the bodily states associated with the emotion in the cortex without actually arousing the body. The as-if pathway can also generate the *cognitive* effects associated with the emotional response via the release of nonspecific neuromodulators from nuclei in the brain stem. A kind of brain shorthand, it is *as if* the body is included in the circuit, even though actual feedback from the body is missing or strongly attenuated. This means that the dynamic resonance and amplification of arousal between the body and brain is missing or greatly decreased. But because *as-if* responses register in the somatosensory cortices of the cortex, they can momentarily feel as strong as a full body response.[15] The as-if pathway has probably become increasingly important in evolution. It also becomes more dominant as people mature.

Several regions of the prefrontal cortex have a role in the secondary emotional pathway (whether the body loop is actual or "as-if"). The ventromedial prefrontal cortex is particularly important (Fig. 5.1, right). A neural hub, or convergence zone, it receives and sends signals to and from most regions of the brain. The ventromedial prefrontal cortex is especially well connected to the amygdala. Both are part of the value-encoding network of the brain, which also includes the orbitofrontal region, the striatum, the anterior cingulate, and the insula, among other regions. This broad network calculates both the real and emotional values of objects and behavioral options. The ventromedial region is specialized to signal the emotional valence of objects by reinstating the actual somatic pattern (or the as-if version) associated with them in the past in the somatosensory cortices.[15]

The ventromedial prefrontal cortex is also closely connected with other areas of the prefrontal cortex. The dorsolateral region is critical for attention, short-term memory, and working memory. It can heighten the secondary emotional response by maintaining thoughts and memories about emotion-provoking objects and behavioral options in an active state. The dorsolateral region is not directly connected to the amygdala, but influences it via the ventromedial prefrontal cortex.

Lesion studies show that if the connections between the amygdala and the ventromedial prefrontal cortex are cut, the secondary fear response may not occur, or it is greatly impoverished.[8] Someone with a brain lesion that interfered with these connections might remember some of the details of the "breakup" in the situation described in the previous section, but he or she would not necessarily react to these memories in an emotional way. Instead those details might appear more like a series of facts.

The absence of an emotional response when seeing a clip of a movie associated with a difficult breakup may well seem a desirable state of affairs. Yet emotion has a prospective or motivational role as well as a reactive one. Being able to learn from one's mistakes only at a rational level turns out to be a severe disadvantage. The work of Antonio Damasio and colleagues has shown that emotion is crucial to adaptive decision making, especially in the personal realm.

## EMOTION AND DECISION MAKING

Neuroscientist and clinician Damasio in the 1990's recounted the story of a patient he called Elliot, whose surgery to remove a tumor cut the connection between the amygdala and the ventromedial part of the prefrontal cortex.[8] After the operation, Elliot was unable to manage his time at work, lost his job, and failed at a succession of new ones. His home life became fraught with problems and his wife eventually left him. At the core of his difficulties was a tendency to make bad decisions and an inability to learn from his mistakes.

Elliot's primary emotional response was intact, his intelligence remained superior, his working memory was fine, and according to a variety of tests, he had appropriate social reasoning ability. But his affect seemed subtly flat. Damasio began to suspect that Elliot's decision-making problems were the result of damage to the secondary emotional pathway.

Describing the behavior of a patient with a similar bilateral ventromedial prefrontal lesion and strangely flat affect, who was asked to decide between two alternative dates for his next appointment, Damasio said:

> *The patient pulled out his appointment book and began consulting the calendar. ... For the better part of a half hour, [he] enumerated reasons for or against each of the two dates: previous engagements, proximity to other engagements, possible meteorological conditions, virtually anything that one could reasonably think about concerning a simple date. [Completely calmly, he walked through] a tiresome cost–benefit analysis, an endless outlining and fruitless comparison of options and possible consequences. It took enormous discipline to listen to all this without pounding the table and telling him to stop, but finally we did tell him, quietly, that he should come the second of the alternative dates. His response was equally calm and prompt. He simply said: "That's fine."*[8 (pp. 193–4)]

## A PERSONAL GLIMPSE OF THE ROLE OF EMOTION IN DECISION MAKING

When I first came across Damasio's description of Elliot and other patients who had experienced bilateral ventromedial prefrontal damage, I could not picture their emotional flatness. Nor could I appreciate, other than at an intellectual level, the connection between emotion and decision making pointed to by Damasio's work. I then had an experience that gave me an inkling of what life might be like in the absence of secondary emotion.

At the end of a particularly long day of writing, unbidden, a weak, flat-sounding voice in my mind told me that when I took up my work the next morning I should concentrate on amplifying a certain point. I gave no credence to this directive. Because of its flatness it had no force, and it quickly faded from my mind.

The next morning when I sat down to my writing, I recognized with gusto that I should begin by amplifying a certain argument. A short while later, I realized with some amazement that this was the very point the strange flat voice had directed me to the evening before! Indeed the emotion, or passion, conveyed the sense that the directive was both valid and important, something the disembodied voice of the evening before had not.

Emotions convey a sense of the importance and reliability of the mental contents to which they are attached. Thoughts stripped of feeling carry little weight. Without emotion, to borrow a phrase from Gertrude Stein, "there is no there, there" to guide decision making.

## THE IOWA GAMBLING TASK

Antoine Bechara,[17] along with Damasio[8] and other colleagues, carried out a series of landmark experiments that confirmed that the decision-making difficulties in patients such as Eliot with bilateral ventromedial prefrontal lesions resulted from a faulty secondary emotional pathway. The researchers measured the level of unconscious fear in patients with bilateral ventromedial prefrontal cortex damage and in normal subjects as they gambled for play money in the laboratory. They followed their galvanic skin response (GSR), which measures changes in skin conductivity from miniscule amounts of sweat generated by the release of stress hormones in response to activation of the amygdala.

Each subject sat in front of four decks of cards and turned over a card from any one of them. Depending on the card they received, they were either rewarded a certain amount of play money or were required to give a certain amount back. They were told that the goal was to win as much money as possible, but not how long the game would continue.

Unbeknownst to the players, the decks were fixed. Two gave high rewards as well as stiff penalties and the other two gave half the reward but lighter penalties. The only way to avoid a net loss over the 100 card turns allowed was to choose consistently from the two low-risk decks. The reward–punishment schedule, which is shown in Fig. 5.2, was varied and complex.

The GSR of normal subjects (recorded for 5 s before each choice) revealed that they began to associate unconscious anticipatory fear with the high-risk decks after picking only a few high-penalty cards. This fear increased with each high-penalty card they chose. Although it remained below awareness, it clearly influenced their behavior; after approximately 30 cards most of them began to prefer the low-risk, low-reward decks. Their preference for these "good decks" grew stronger as the game progressed.

After the 20th card, the subjects were asked what they knew about the different decks, and they were subsequently asked the same question after every 10 cards. About 70% of the normal subjects expressed a hunch that two of the decks were better or safer *before* they understood at a conscious level that the other two were stacked against accumulating money.

In contrast, the brain-damaged subjects continued to prefer the high-risk, high-reward decks throughout the experiment. Their GSRs were nearly flat before choosing from the high-risk decks, indicating that they were not experiencing anticipatory unconscious fear. Moreover none reported having a hunch that some of the decks might be safer or better than others.

Yet their GSR to picking a high-penalty card seemed intact. It was similar to low-responding normal subjects. Moreover they stopped choosing from the risky decks for a few turns afterward. This is because their primary fear pathway, which goes directly to the amygdala, was still functional. (Patients with bilateral *amygdala damage*, in contrast, have flat GSRs to picking a high-penalty card as well as to choosing from the risky decks.[18])

Half of the patients with ventromedial lesions eventually came to recognize at a conscious level that the only way to accumulate money was to choose consistently from the two low-penalty decks, although it took them considerably longer than the normal subjects. Even after they realized how the decks had been constructed, strikingly, they continued to choose cards from the riskier desks. Likewise, their GSRs still did not register fear before they picked from the "bad" decks. Asked why they continued to choose from them, they commented that it was more interesting this way or that the rules might change at any time.[8]

**Original Task (ABCD)**

| Deck | 1 | 2 | 3 | 4 | 5 | 6 | 7 | 8 | 9 | 10 | 11 | 12 | 13 | 14 | 15 | 16 | 17 | 18 | 19 | 20 | 21 | 22 | 23 | 24 | 25 | 26 | 27 | 28 | 29 | 30 | 31 | 32 | 33 | 34 | 35 | 36 | 37 | 38 | 39 | 40 |
|---|---|---|---|---|---|---|---|---|---|---|---|---|---|---|---|---|---|---|---|---|---|---|---|---|---|---|---|---|---|---|---|---|---|---|---|---|---|---|---|---|
| A(+100) | | | -150 | | -300 | | -200 | | -250 | -350 | | -350 | | -250 | -200 | | -300 | -150 | | | | -300 | | -350 | | -200 | -250 | -150 | | | -350 | -200 | -250 | | | | -150 | -300 | | |
| B(+100) | 0 | 0 | | 0 | | 0 | | | -1250 | | 0 | 0 | | -1250 | | 0 | 0 | | | | -1250 | | 0 | 0 | | 0 | | 0 | | | 0 | -1250 | | 0 | | 0 | | 0 | | |
| C(+50) | | | -50 | | -50 | | -50 | | -50 | -50 | | -25 | -75 | | | | -25 | -75 | | -50 | | | | -50 | -25 | -50 | | | -75 | -50 | | | | -25 | -25 | | -75 | | -50 | -75 |
| D(+50) | 0 | | 0 | | | 0 | 0 | | | -250 | 0 | 0 | | | 0 | | | | 0 | -250 | | 0 | | | 0 | 0 | 0 | | -250 | | | 0 | | 0 | -250 | 0 | | | 0 | |

**Figure 5.2.** ***Gain and loss schedule for cards from each deck in the original version of the Iowa Gambling Task.*** Deck A and B give players $100 every time they choose a card from them but also give steep penalties (the amount of money they have to give back). Negative numbers in a box indicate the penalty for turning over that card from the deck. Empty boxes or those with 0 in them indicate no penalty for that card. Decks C and D give only $50 for each card turned over from them, but they give lower penalties as well. The first time players choose from any deck, they get the first card in the deck; the second time, the second card from it, etc. The game is over after 100 card turns. The only way to come out ahead is to begin choosing consistently from the lower-reward, lower-penalty decks. *Reprinted from Bechara A, Tranel D, Damasio H. Characterization of the decision-making deficit of patients with ventromedial prefrontal cortex lesions. Brain 2000;123:2189–202, with permission from Oxford University Press.*

The disadvantageous choices of the brain-damaged subjects could reflect a heightened sensitivity to reward, a decreased sensitivity to punishment, or a general insensitivity to long-term consequences that are either punishing or rewarding. To decide among these alternatives, the researchers used a reversed version of the gambling task.[19] In this version, the advantageous decks give twice the punishment as the disadvantageous ones but much higher rewards. The successful strategy here is to choose consistently from the high-punishment but higher-reward decks.

Normal subjects again came to prefer the decks that would lead to long-term gain even though it meant favoring those that gave higher punishment. Patients with bilateral ventromedial lesions did not make advantageous choices in the reversed gambling task either, which was consistent with the absence of anticipatory GSR on the initial gambling task. Thus neither heightened sensitivity to reward, which might have accounted for their response on the original version of the test, nor decreased sensitivity to punishment, which might have accounted for their response on the alternative version, can explain their choices. Instead their lesions left them focused on the short term, or "myopic" to future consequences. This short term focus accounts for their poor decision making. In contrast, many problematic gamblers show heightened sensitivity to reward on the original tests.[20]

The lesions of the brain-damaged subjects prevented them from having the anticipatory emotional signals based on past experiences that normal subjects had. Because the connections between the amygdala and prefrontal cortex were severed, the brain-damaged subjects did not generate the somatic responses, or changes in bodily state, that in normal subjects were associated with anticipation of the consequences of choosing from the risky deck. These bodily signals, which become juxtaposed with images of contemplated action in the prefrontal cortex, were named *somatic markers* by Damasio.

It is tempting to use the term "gut feeling" for somatic markers. However Damasio's term is broader than a literal understanding of the phrase "gut feeling" would suggest. Somatic markers include an infinite variety of different bodily sensations that can act as a signaling and biasing mechanism. Also, although they sometimes register as a "feeling" in consciousness or its fringe, they can function completely below awareness.

Likewise, *choosing* in these tasks does not necessarily mean a conscious choice. Studies by Soon and colleagues (see section Who's in Charge? in Chapter 1),[21] building on earlier work of Libet,[22] suggested that seemingly voluntary choices can demonstrate characteristic neural signals in the prefrontal cortex as much as 10 s before the subject is consciously awareness of the decision. The biasing action of somatic markers in the gambling experiment may occur below awareness in response to unconsciously contemplated action well before subjects feel the conscious urge to turn a card from one deck or the other, especially after the initial phase.

Bechara and Damasio proposed both a subcortical and a cortical mechanism for the adaptive effect of somatic markers on decision making, in accord with their unconscious and conscious influences.[15] In the subcortical mechanism, release of dopamine from the mesolimbic dopamine system, where it functions as a neurotransmitter, automatically biases decision making below awareness toward longer-term rewards based on past experience. In the gambling task, this can account for the growing preference of normal subjects for the advantageous deck in the early prehunch stage. Mesolimbic dopamine release is particularly sensitive to unpredictable rewards.[23]

The cortical mechanism to account for anticipatory secondary emotional responses, in contrast, generates potentially conscious feeling states (feelings of "goodness" or "badness") associated with the contemplated behavior. It can account for the feeling subjects have during the hunch period that two of the decks are better or safer before they have figured out why. These responses involve the anterior insula, which is especially active in emotion-tinged cognitive evaluations involving risk, aversion, uncertainty, and complexity, all of which apply to the gambling task.[24]

The anticipatory GSRs of the normal subjects in both the original and reversed tasks were a response to autonomic nervous system activation. This suggests that they had a bodily rather than an as-if secondary emotional response. The researchers proposed that decisions made with uncertainty with respect to the odds—as they argued holds throughout the gambling experiment—are more likely to engage the bodily loop.[15] In contrast, when the odds are certain and the response contingencies more straightforward, and often more limited, they can be represented by central nervous system activity as *as-if feelings*. More ambiguous situations involving more response options seem to require the more subtle differential discriminations that are possible in the body loop.

## THE IOWA GAMBLING EXPERIMENTS AND INTUITION

The claim that normal subjects begin to prefer the advantageous decks before they have actual conscious knowledge that they are a better choice has, not surprisingly, drawn considerable experimental scrutiny. The results of two of these studies are particularly interesting.[25,26]

Taken together, they suggest that the extent to which knowledge is available to awareness can be altered by questioning subjects much more extensively after every 10 cards about their cumulative experience of each deck and their expectation about them in the future. This forces subjects to reflect on perhaps even subtle aspects of their recent experience of the different decks. It also likely sensitizes them to their ongoing experience of them as they make their subsequent choices, which otherwise might remain below awareness. These results lend support to the view presented in Chapter 2 that the division between unconscious and conscious knowledge lies on a continuum and is sensitive to a variety of different factors.

Experiments with a version of the gambling task specifically designed for rats also provide evidence consistent with the unconscious status of the early choices in the gambling experiments. They showed that rats also can choose advantageously in situations that require the complex integration of differing probabilities along with varying magnitudes of rewards and punishment.[27] Rats are unlikely to make decisions consciously, which lends credence to the idea that humans can make complex behavioral evaluations below awareness. The capacity for sophisticated unconscious decision making apparently is part of our evolutionary inheritance.

In humans, the extensive development of the prefrontal cortex confers the ability to predict the consequences of behavior further into the future than lower mammals, as well as more steps ahead and based on more abstract considerations. These extended capabilities all can function below awareness to enhance decision making. They appear to depend on the more anterior part of the prefrontal cortex, which has preferentially evolved in primates.[28] For example, patients with bilateral *anterior* ventromedial lesions tend to be selectively impaired in bodily representations and therefore decision making that involves consequences in the far future rather than the near future.

The gambling experiments highlight the important links among the body, emotion, and intuition. Damasio's patients with a damaged frontal lobe who cannot harness somatic markers to anticipate what is likely to happen in recurring situations experience an inability to use intuition in decision making involving well-being. Together, these studies provide strong biological support for some basic aspects of intuition. They demonstrate that emotion-derived signals can successfully replace conscious reasoning in certain decision-making situations. They can signal choices that are risky *well before* the conscious mind can recognize them.

This emotional knowledge then can ground and reinforce any subsequent conscious reasoning and choice. Indeed without these intuitive bodily signals, conscious knowledge may be of limited use—it has no oomph. Thus the brain-damaged subjects continued to sample from the "bad decks" in the gambling task *even after* they caught on that the decks were a poor choice.

Yet anticipatory bodily signals are not always a reliable guide for current action for the very reason that they are based on past experience. Somatic markers can sometimes color current choices in disadvantageous ways. In real life, the odds sometimes *do* change without warning in response to a variety of different factors. Because patients with ventromedial lesions are less affected by recent gains or losses, in certain investment situations they can come out ahead.[29]

Both genetic and environment factors influence decision-making proclivities below awareness.[18,30,31] Learned decision-making proclivities that proved adaptive in difficult childhood environments may become fixed into perceptual habits that skew perception later in life in ways that lead to suboptimal choices, and this is especially so especially if they overlap with genetic tendencies.[32] Emotional contamination from unrelated events can also influence decision making. A recent study showed that subjecting people to an annoying situation before giving them the gambling task made them avoid decks that offered more frequent penalties over ones that offered larger infrequent ones, although both gave the same payout.[33]

## DEEP INTEGRATION OF EMOTION AND COGNITION

In the primary pathway, the emotional response is activated by sensory information prior to cortical processing. In the secondary emotional pathway, the emotional response is activated by sensory information after it has been fully processed by the sensory cortices. Yet emotion-evoking visual stimuli can influence their own *cortical* processing in a selective fashion more quickly than the time it takes for them to reach the last stage of processing in the visual cortex.[34,35] Brain imaging studies point to at least one, and likely multiple additional cortical pathways that integrate emotion and cognition/perception from early on in the sensory stream.[36,37]

One, the magnocellular pathway, goes through the first stage of visual processing.[36] Its initial target is the *orbitofrontal area*, which is partially coextensive with the ventromedial prefrontal cortex. Just behind the eye sockets, the orbitofrontal area, as noted before, is an important part of the value-encoding network of the brain that calculates the biological value of objects. Such a fast pathway is consistent with rapid feed-forward activity seen in imaging studies that goes from early visual processing in the back of the brain to the front of the brain.[38]

Moshe Bar and colleagues[36] showed that the fast magnocellular visual pathway carries sketchy or low-frequency "gist" information about the focal stimulus, as well as some context information attached that is to it, to the orbitofrontal area. They argued that this rapid, gist-like pathway provides a guess, or prediction, about the identity of the object and also evokes bodily feeling and emotions in response to this prediction. The information is then fed back into the slower, more precise visual pathway, where it guides subsequent processing by, among other things, limiting choices and resolving ambiguity.[39] This would allow emotion an early top-down role in visual recognition. Other bidirectional pathways also seem to connect the early visual stream and various regions of the brain important to evaluating sensory information. Together they suggests that visual input and emotion are integrated at many intersecting levels. Indeed Pessoa[40] argued that emotional information influences cognition in multiple waves.

A fast, gist-like pathway such as the magnocellular may well have a place in some experiences of intuition. It can explain certain hunches. It would allow the brain to know in a quick and dirty way whether something appears coherent, what it might be, as well as how we might feel about it before the brain actually recognizes what it is.[41] However such a pathway cannot account for all intuition. High-level intuition is neither necessarily quick nor gist-like. It can encode startling complexity and detail. Moreover it often develops over time.

At a more fundamental level, a rapid cortical pathway responsive to emotional influences offers a mechanism for the brain to guess the probable importance or value of objects or events in order to determine their appropriate level of processing. As Freeman's[42] work described in Chapter 4 stressed, there is too much information in the environment for the brain to process all of it. It must select what is likely to be most important. A fast pathway or pathways would allow the brain to signal the emotional import of an object or event to the organism back into the more detailed sensory-processing stream. This provides a potential way out of the circular conundrum of the brain having to know what something is, and thus its importance, before it determines how much processing it should undergo. An early emotional assessment could selectively enhance the depth of processing of a low-level stimulus related to current or long-term needs and goals, which might otherwise might be lost. Boosting the strength of certain low-level inputs because of their potential importance can sometimes be critical to the significant rearrangement of the cognitive landscape that characterizes breakthrough intuitions.

The existence of the fast pathway is one of a number of pieces of evidence that have led some researchers to propose that the older distinction between emotional and cognitive regions of the brain is outdated.[40] Instead they suggest that all parts of the brain are both emotional and cognitive, although some may be more one than the other. In fact, work from the judgment and decision-making tradition (a former bastion of the cognitively grounded perspective) has suggested the possibility that all risk–benefit analysis may be based on emotion, and no less rigorous for it.[43]

This increasing blurring of the traditional boundaries between emotion and cognition is also consistent with the view that all brain regions are directly or indirectly interconnected in networks.[40] As the last chapter stressed, the brain functions as an interconnected whole. As a complex dynamical system, its response in each instant is a complex cooperative effort. Moreover, it likely encompasses considerable redundancy. As a result, it has become increasingly difficult to attribute any response to only one anatomical structure or pathway. Also the connection strength or functional connectivity among different regions can change from instant to instant, no doubt in response to subtle as well as larger alterations in conditions. The amygdala, the orbitofrontal area, and the insula because of their direct connectivity to so many other brain areas form especially important nodes within the interconnected brain.

## EMOTION IN MORE IMPERSONAL COGNITION

As the previous section suggested, the entwining of emotion and cognition is critical to every level of intelligent functioning. The gambling experiments also provide a biological framework for understanding the role of emotion in more impersonal cognition. Emotion helps determine the prominence, or saliency, as well as the valence of various objects, events, and patterns. Damasio proposed that somatic markers can also function at a covert level "to highlight in the form of an attentional mechanism, certain components over others, and to control in effect, the go, stop and turn signals necessary for some aspect of the decision-making and planning in non-personal, nonsocial domains."[8(p190)]

In a similar vein, he pointed out that unconscious emotion seems to streamline intellectual life by bringing to consciousness only what would benefit most from conscious consideration.[8] In agreement with an idea first proposed by Poincaré in the 19th century, he argues that intuition orients the conscious mind by limiting the universe of possibilities and presenting only the most promising mental connections to awareness.[8] Working through unconscious emotion, intuition can aid rational thought by offering the implicit suggestion "Start your thinking here."

However these veiled cognitive evaluations do not depend exclusively on past experience. Learned contingencies are at best only one of the currents informing the intellectual judgments that somatic markers can report. Many of the deepest intuitions, whether they are personal or intellectual, require a fresh start that is to a greater or lesser extent scrubbed of what has worked in the past. Chapter 8 on metacognition explores in some detail the functioning of the emotionally tinged signals that monitor, evaluate, and report breakthrough understanding below awareness or on its fringe as it occurs.

## PUTTING IT TOGETHER

Emotional responses are grounded in genetic survival. They are nature's way of tethering organisms to the world at a biological level. Our innate bodily needs and responses in conjunction with those that develop as we interact with the environment generate a psychological terrain that is sculpted by what we need and want, as well as what we fear. The more subtle characteristics of this landscape remain largely hidden. These "values," both hidden and overt, provide the motivation for our intuitions, as they do for all our cognition and behavior.

The peaks and valleys of this terrain make us especially responsive to relevant knowledge from the environment. Somatic markers allow us to assimilate this knowledge automatically in an anticipatory way into unconscious decision making. Frequently the adaptive behavior just happens, or the answer to the not yet asked question just appears in awareness at the right time. At other times, this emotional landscape provides the intuitive assumptions from which we commence conscious reasoning. However, because the automatic urgings that result from this landscape are grounded in what happened to further well-being or survival in the past, sometimes they may be limited in scope or even counterproductive. Often deeper intuition represents the spontaneous reorganization of unconscious perception, which in the end might integrate better with other deeper goals, as well as some of the other things we know.

## REFERENCES


1. Darwin C. *The expression of the emotions in man and animals*. 3rd ed. New York: Harper Collins, Oxford University Press; 1998.
2. Lettvin JY, Manturana HR, McCulloch WS, Pitts WH. What the frog's eye tells the frog's brain. In: Maturana HR, McCulloch WS, editors. *Embodiment of mind*. Cambridge: MIT Press; 1965.
3. LeDoux J. Fear conditioning can be demonstrated even in creatures as lowly as snails *The emotional brain*. New York: Touchdown Touchstone; 1999. p. 147.
4. Emotions can be viewed as a subset of drives. Drives initially cause arousal from within—for example, hunger and thirst—whereas emotions do so from without. However the distinction disappears quickly because of the conditioned nature of the arousal response. Objects to which drives attach themselves also become emotional inducers and cause arousal from outside.
5. Bowlby J. *Attachment and loss. Attachment*, vol. 1. New York: Basic Books; 1982.
6. Marcoa EM, Llorentec R, López-Gallardo M. The maternal deprivation animal model revisited. *Neurosci Biobehav Rev* 2015;**51**:151–63.
7. Schore AN. The effect of early relational trauma on right brain development, affect regulation and infant mental health. *Infant Ment Health J* 2001;**22**:201–69.
8. Damasio A. *Descartes' error: emotion, reason and the human brain*. New York: G.P. Putnam's Sons; 1994.
9. I sometimes think of this as the "foxhole phenomenon." For example, when the immediate environment is filled with chaos, sometimes there is nothing more soothing than focusing on the limiting confines of the computer screen.
10. Damasio A. *The feeling of what happens: body and emotion in the making of consciousness*. New York: Harcourt; 1999.
11. LeDoux J. *The emotional brain*. New York: Touchstone; 1998. This would be based on information arriving either directly from the thalamus or after feature extraction from early visual processing.
12. James W. *Mind* 1884;**9**:188–205.
13. Critchley HD, Wiens S, Rotshtein PA, Ohman A, Dolan RJ. Neural systems supporting interoceptive awareness. *Nat Neurosci* 2004;**7**:189–95.
14. Uddin LC. Salience processing and insular cortical function and dysfunction. *Nat Rev Neurosci* 2015;**16**:55–61.
15. Bechara A, Damasio AR. The somatic marker hypothesis. *Games Econ Behav* 2005;**52**:336–72.
16. Jiang Y, Costello P, Fang F, Huang M, He S. A gender- and sexual-orientation dependent spatial attentional effect of invisible images. *Proc Natl Acad Sci USA* 2006;**103**:1748–52.
17. Bechara A, Damasio H, Tranel D, Damasio AR. Deciding advantageously before knowing the advantageous strategy. *Science* 1997;**275**:1293–5. In another study, a few normal subjects chose more cards from the bad decks and also had higher SCR to the good than the bad decks. Bechara A, Damasio H, Damasio A, Lee G. Different contributions of the human amygdala and ventromedial prefrontal cortex to decision-making. *Journal of Neuroscience* 1999;**19**:5373–481.
18. Bechara A, Damasio H, Damasio AR, Lee GP. Different contributions of the human amygdala and ventromedial prefrontal cortex to decision-making. *J Neurosci* 1999;**19**:5373–481.
19. Bechara A, Tranel D, Damasio H. Characterization of the decision-making deficit of patients with ventromedial prefrontal cortex lesions. *Brain* 2000;**123**:2189–202.
20. Brevers D, Bechara A, Cleeremans A, Noël X. Iowa gambling task (IGT): twenty years after-gambling disorder and IGT. *Front Psychol* 2013;**4**:665. https://doi.org/10.3389/fpsyg.2013.00665.

21. Soon HJ, Brass M, Heinze H-J, Haynes J-D. Unconscious determinants of free decisions in the human brain. *Nat Neurosci* 2008;**11**:543–5.
22. Libet B. Unconscious cerebral initiative and the role of conscious will in voluntary action. *Behav Brain Sci* 1985;**8**:529–66.
23. St Onge JR, Floresco SB. Dopaminergic modulation of risk-based decision making. *Neuropsychopharmacology* 2009;**34**:681–97.
24. Singer T, Critchley HD, Preuschoff K. A common role of insula in feelings of empathy and uncertainty. *Trends Cogn Sci* 2009;**13**:330–40.
25. Maia TV, McClelland JL. A reexamination of the evidence for the somatic marker hypothesis: what participants really know in the Iowa gambling task. *Proc Natl Acad Sci USA* 2004;**101**:16075–80.
26. Persaud N, McLeod P, Cowey A. Post-decision wagering objectively measures awareness. *Nat Neurosci* 2007;**10**:257–61.
27. Zeeb FD, Robbins TW, Winstanley CA. Serotonergic and dopaminergic modulation of gambling behavior as assessed using a novel rat gambling task. *Neuropsychopharmacology* 2009;**34**:1–15.
28. Koritzky G, He Q, Xue G, Wong S, Xiao L, Bechara A. Processing of time within the prefrontal cortex: recent time engages posterior areas whereas distant time engages anterior area. *NeuroImage* 2013;**15**:280–6.
29. Shia B, Lowenstein G, Bechara A, Damasio H, Damasio AR. Investment and the negative side of emotion. *Psychol Sci* 2005;**16**:435–9.
30. Bechara A, Dolan S, Hindes A. Decision-making and addiction (part II): myopia for the future or hypersensitivity to reward? *Neuropsychologia* 2002;**40**:1690–705.
31. Bechara A, Damasio H, Damasio AR. Emotion, decision making and the orbital frontal cortex. *Cereb Cortex* 2000;**10**:295–307.
32. Bishop SJ. Neural mechanisms underlying selective attention to threat. *Ann NY Acad Sci* 2008;**1129**:141–52.
33. Yechiam E, Telpaz A, Krupenia S, Rafaeli A. Unhappiness intensifies the avoidance of frequent losses while happiness overcomes it. *Front Psychol* November 2, 2016;**7**:1703. eCollection. 2016.
34. Eimer M, Holmes A. An ERP study on the time course of emotional face processing. *NeuroReport* 2002;**13**:427–31.
35. Kawasaki H, Adolphs R, Kaufman O, Damasio H, Damasio AR, et al. Single-neuron responses to emotional visual stimuli recorded in human ventral prefrontal cortex. *Nat Neurosci* 2001;**4**:15–6.
36. Bar M, Kassam KS, Ghuman AS, et al. Top-down facilitation of visual recognition. *Proc Natl Acad Sci USA* 2006;**103**:449–54.
37. Rudrauf D, David O, Lachaux J-PL, et al. Rapid interaction between the ventral visual stream and emotion-related structures rely on a two-pathway architecture. *J Neurosci* 2008;**28**:2803–973.
38. Lamme VAF, Roelfsema RP. The distinct modes of vision offered by feedforward and recurrent processing. *Trends Neurosci* 2000;**23**:571–9.
39. Barrett F, Bar M. Say it with feeling: affective predictions during object perception. *Philos Trans Roy Soc B* 2009;**364**:1325–34.
40. Pessoa L. On the relationship between emotion and cognition. *Nat Rev Neurosci* 2008;**9**:148–58.
41. Volz KG, Rübsamen R, von Cramon DY. Cortical regions activated by the subjective sense of perceptual coherence of environmental sounds: a proposal for a neuroscience of intuition. *Cogn Affect Behav Neurosci* 2008;**8**:318–28.
42. Freeman WJ. *How brains make up their minds*. London: Weiden & Nicolson; 1999.
43. Quartz SR. Reason, emotion and decision-making: risk and reward computation with feeling. *Trends Cogn Sci* 2009;**13**:209–15.

# CHAPTER 6

# Mental Imagery, Imagination, and Intuition

## Contents



Empirical observation is only one step in scientific endeavor. The deeper goal is to move behind measurements to a more basic and complete level of comprehension. Intuition as it comes through mental imagery is often essential to achieving this fundamental knowledge. This chapter explores how mental imagery supports intuition and also examines its potential role in revealing profound scientific understanding.

Intuitive imagery has two different functions in science. The second, more acknowledged one is to make scientific insight readily accessible to others. The first, which often goes unrecognized, is to bring unconscious knowledge to awareness.

A variety of different sensory modalities can bring unconscious insight to awareness, and this is as true in science as it is in any other human endeavor. Intuitive images may be as completely structured and clear as the periodic table that came to Mendeleev in a dream,[1] or they may be as vague and difficult to pin down as subtle muscular sensations that require considerable reflection and interpretation. Several sensory modes may participate together.

Even in mathematics, the most abstract and rigorous discipline, a number of different kinds of subjective mental imagery can bring mental


*Understanding Intuition*
ISBN 978-0-12-814108-3
https://doi.org/10.1016/B978-0-12-814108-3.00006-3

contents to consciousness. Jacques Hadamard in his 1945 book, *The Mathematician's Mind: The Psychology of Invention in the Mathematical Field*, reported on the responses of his colleagues' to a survey he circulated asking what kinds of mental images carried their thoughts to awareness.[2] Einstein responded that his were muscular as well as visual. Another reported that his mind used a word or letter to symbolize a whole line of thought. Another said nonsense words, and another, auditory images.

Hadamard was member of a school of mathematics called *intuitionism*. He derived proofs using mental imagery and only later formalized them with equations. To illustrate, he offered an intuitive proof that the series of prime numbers (numbers that have no factors) is infinite.[2(pp77–78)] His mental images appear side by side with each step of the formal classical proof.

| Hadamard's Mental Picture | Steps in the Proof |
|---|---|
| 1. I see a confused mass. | I consider all primes from 2 to 11, say 2, 3, 5, 7, 11 |
| 2. N being a rather large number, imagine a point rather remote from the confused mass. | I form their product I $2 \times 3 \times 5 \times 7 \times 11$ |
| 3. I see a second point a little beyond the first. | I increase that product by, say, N plus 1 |
| 4. I see a place somewhere between the confused mass and the first point. | That number, if not a prime, must admit of a prime divisor, which is the required number. |

Hadamard stresses that the purpose of this "strange and cloudy imagery" is to give the proof an all-over shape, to allow him to conceive it as a whole:

> *Certainly, it is not meant to remind me of any property of divisibility, prime numbers and so on. ... I need it in order to have a simultaneous view of the argument—in short to achieve that synthesis ... and give the problem its physiognomy. ... Indeed, every mathematical research compels me to build such a schema.*[2(p77)]

Although a variety of different modes of imagery can bring insight to awareness, complex ideas in science are often made accessible in visual form, especially to others. Words and equations require sequential, or stepwise, understanding. In contrast, visual imagery can represent multiple relationships simultaneously. This allows all of the different strands as well as their interactions to be grasped as a whole.

Ideas in science that can be rendered accessible visually are often labeled *intuitive*. The essence of quantum mechanics *could not* be captured visually and therefore was considered *unintuitive*. Of course, not every visual rendition of information in science necessarily supports intuition. Graphs, for example, reflect a largely mechanical rendering of the data in compact visual form. Nonetheless, because they allow a large amount of data to be pictured at one time, they sometimes help initiate intuitions. They may reveal patterns that allow the viewer to integrate their relevant unconscious as well as conscious information and as a result yield new and sometimes breakthrough understanding.

On occasion, the two different functions for intuitive imagery in science converge: bringing unconscious insight to awareness and communicating abstractions in readily comprehensible form. Most often working through visual images, intuition can go beyond the data to make apparent the underlying relationships and laws our senses cannot register and our instruments cannot record. Because of this, discussions of the role of intuition in science tend to focus largely on the ability of certain scientists to conceptualize abstractions visually.

Physicist Richard Feynman, who studied quantum electrodynamics (the transformations elementary particles undergo when they collide) was celebrated for his intuition. Before Feynman, all problems in the field were solved with equations. Yet the underlying dynamics of particle interactions came to Feynman in diagrams that condensed a large amount of essential data and presented it simultaneously (Fig. 6.1).[3] These visualizations allowed him to streamline the calculations involved in solving complex problems,

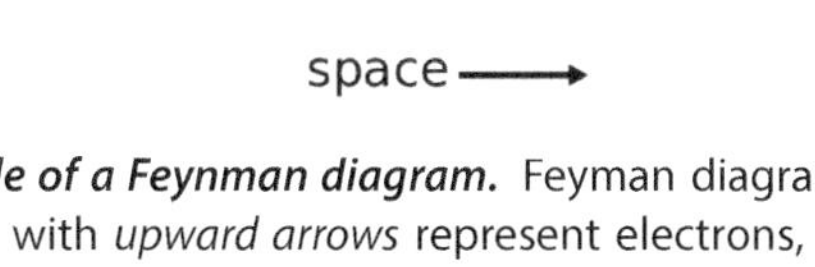


**Figure 6.1** ***An example of a Feynman diagram.*** Feyman diagrams are read from bottom to top. *Solid lines* with *upward arrows* represent electrons, *solid lines* with *arrows facing downward* represent positrons, or as Feynman realized, electrons moving backwards in time. The *wavy lines* represent photons. An interaction or event is represented by a dot. Each of these elements represents a term in an equation. The diagram depicts an electron and a photon colliding and forming a positron and a photon. Successively rotating the diagram 90 degrees reveals all possible interactions between an electron, a positron, and two photons.

greatly extending the reach of the field. They also suggested a radically new way of understanding particle interactions. Feynman diagrams are now widely used to describe the behavior of subatomic particles, and they have also infiltrated much of theoretical physics.[4]

## WHAT IS MENTAL IMAGERY?

Let's step back momentarily from intuitive images and consider mental imagery more generally. Mental images can be divided into two broad general classes. Some transmit actual sensory information from either the external world, such as a car speeding by as we wait to cross a busy street, or the internal world, such as a stomachache. Others we generate only mentally: for instance, the vivid midafternoon reverie of enjoying a relaxing dinner at a favorite restaurant with a friend.

At one level of description, the brain processes sensory signals in a complex sequence of steps. Eventually it forms higher-order neural patterns of those that are meaningful, recorded in the tendency of certain neurons to fire together. These neural patterns come to correspond to specific aspects of the world. Among other things, they bind together different characteristics of an object, such as its shape, color, and sound, as well as perhaps its tactile feel. Thus we come to see the waterfall at the same time that we hear its roar, smell the freshness of its spray, and feel its dampness on our skin.

Information coming from the senses corresponds in some way to the external or internal environment. However by no means do we perceive what is actually there. Human sensory systems, like those of every species, have been honed by evolution to transmit primarily the kind of information that proved essential to survival during the course of evolution. Unlike dogs, our sense of smell is weak and our hearing excludes the ultrasonic. Unlike bees, we cannot see ultraviolet radiation, and in contrast to certain snakes, we cannot see in the infrared range.

Moreover, we do not necessarily *perceive* even what our senses record. The brain ignores some sensory inputs in favor of others that carry apparently more important information. When we are hungry, food-related billboards and signs are more likely to capture our attention than after a meal.[5] Also past experience conditions each of us to be more sensitive to certain aspects of the world than others. Our sensory machinery is subservient to survival and well-being or, more broadly, to meaning making, as Walter Freeman's work in the second half of Chapter 4 emphasizes.[6]

Even in the absence of current sensory input from the external world, we can recall information previously stored in the tendencies of different complexes of neurons to fire together in patterns. We can thus recreate a reasonable mental facsimile of what we have seen, heard, or touched. These internal images seem to use much the same processing apparatus as does direct sensory input.[7,8] However they tend to be weaker, which keeps us from confusing mental images generated internally with current perceptions of the external world.

As we interact with the external world, we each develop our own personal inner model of it that conforms to our experiences. Yet our model is not handed to us passively. It is formed and refined as we actively engage with the environment over time to fulfill our needs. The model also comes to determine what we can easily recognize, as well as our expectation of what we will see. Just as the members of the proverbial group of blind people feeling an elephant all develop different ideas about the animal depending on the part of it they happen to explore, so too our own activity is critical to our internal model of the world.

Nonetheless *within broad limits*, all humans tend toward similar mental models of the sensory world. The mental images that bring sensory information to consciousness are *reasonably* alike for everyone. We all start with similar sensory and motor equipment, are subject to similar physiological drives, and interact with an external world that is similar in its broad outlines. Cultural factors such as language and education tend to strengthen and extend this mental homogeneity. Because our mental images work so well, they are self-reinforcing. We never need to become aware that, even when they are culturally consensual, they are ultimately no more than a shared mental model.

## INTERNAL IMAGERY, IMAGINATION, AND INTUITION

Imagination provides a critical link between intuition and mental imagery. My dictionary reveals that in addition to what is unreal or fanciful, imagination refers to all mental images of anything not currently present to the senses.[9] In fact, the words *imagination* and *image* are from the same root. According to this view, imagination encompasses *all* internal imagery, including all images of past sensory experiences retained in memory.

Imagination can be seen as much more a part of ordinary mental functioning than we generally suppose. To identify a small animal by the side of the road on a dark night as a raccoon, rather than a large cat, small dog, or a skunk, the mind implicitly connects it to other encounters with small animals of this size with similar characteristics.[6,10,11] If we have had a number of experiences with raccoons, the verbal label "raccoon" or another mental equivalent may just pop into the mind.

Yet even here the mental experience consists of more than just a categorization process, although in most cases everything but the verbal label or its equivalent remains in the background of experience. It is also a *meaning-rendering* process in the full sense of the phrase. Connecting current experience with past experience of raccoons requires imagination in the dictionary sense of the word—of something not present to the senses. Along with past experiences of raccoons, it includes the activation of the salient emotional feelings we might associate with them: perhaps the pleasures of summer camping, and/or a mixture of admiration and resentment at their ability to get into several generations of garbage cans designed to be raccoon-resistant. This is what makes it our own knowledge and conveys the rich sense of personal meaning below awareness or at its fringe with which we all endow our sensory experience.

An exhibit a long time ago at the Exploratorium, an interactive science museum in San Francisco, first brought home to me this level of imagination buried in the background of experience. In a section called "The Mind," one of the displays was a low toilet with a drinking spigot on top of one side of its water tank. A sign invited visitors to drink and assured them it was sanitary. Nonetheless, I suspect that, like me, most found their imagination in overdrive as they bent down to have a sip.

The layer of personal yet shared imagination and inner meaning this experience highlights is part of what gives poetry its broad appeal. (Of course, toilets, never mind drinking from them, are not often the subject of poems.) Poetry uses intersecting webs of shared associations that are largely experienced below awareness to create its fertile and resonant sense of meaning.

This broader view of imagination provides a deeper understanding of the role of intuitive images, one linking science and poetry. Intuitive knowing, which is unconscious knowing, requires imagination, or the process of forming images, to enter consciousness. This formulation separates the process of intuition into two parts: the first is unconscious knowing, and the second is imagination, which bridges the unconscious and conscious mind through mental imagery.

This separation is especially useful when thinking about breakthrough understanding in which past experience provides no ready pattern, schema, or verbal label to carry some of the fullness of unconscious insight to awareness. Imagination forges the specific images, whether realistic or fanciful, with all of their potential subtlety, that bring unconscious knowing to awareness. For example, Barbara McClintock could *see* the chromosomes turning on and off in her mind's eye as she looked at her slides under the microscope,[12] and Kekulé perceived the resonant ring structure of benzene (the basis of organic chemistry) in a waking reverie of snake-like molecules dancing before his eyes as he sat by the fire[13]:

> *I turned the chair to face the fireplace and slipped into a languorous state. Again atoms fluttered before my eyes. Smaller groups stayed mostly in the background this time. My mind's eye, sharpened by repeated visions of this sort, now distinguished larger figures in manifold shapes. Long rows, frequently linked more densely; everything in motion, winding and turning like snakes. And lo, what was that? One of the snakes grabbed its own tail and the image whirled mockingly before my eyes. I came to my senses as though struck by lightning. ...*

## IMAGINATION IN SCIENCE

The collective scientific endeavor can be seen as an extension of the basic human mental activity of *image-ing* or *imagining* the world. Imagination as word is generally used (meaning the formation of *novel* mental images) functions at two related levels in science. It helps scientists to develop new protocols and instruments that can expand or refine past observations, as well as identify new phenomena. In doing so, imagination fine-tunes as well as extends the collective mental model of the natural world.

Yet imagination is not limited to helping scientists make more precise observations or describing new phenomena. It also allows them to develop theories that characterize the essential relationships underlying past and present observations. These theories give coherence to experimental findings and often connect a wide variety of seemingly diverse phenomena. (Theories, in turn, have a critical role in guiding what measurements are made.[14]) Imagination allows scientists continually to deepen and reconfigure the collective model of the world as they seek to uncover and understand its underlying regularities.

These two broad functions of imagination correspond to two separate yet richly interconnected pathways to scientific inquiry, the empirical and the intuitive. On the empirical path, scientists observe the natural world at an increasingly finer level, or perhaps at an increasingly larger scale, over time. This most often requires active probing, and the instruments and experimental paradigms used may require considerable imagination.

On the intuitive path, the emphasis is instead on creating the imaginative underlying theories and images that give coherence to experimental findings. It is not isolated from the empirical path because subsidiary paths potentially link them at every point. Yet the focus on the intuitive path is different from the focus on the empirical one, although sometimes only subtly so.

## CARL JUNG: SENSATE VERSUS INTUITIVE PERCEPTION

Carl Jung's theory of psychological types is helpful for capturing the difference between these two approaches to scientific inquiry.[15] Jung's writing is not grounded in scientific evidence, yet his insights into the intuitive mind are unusually penetrating. He himself strongly favored intuitive perception and cognition; he wrote in part from inside the experience. Attempts to understand intuition *as a whole* that do not take Jung's work into account are likely to be improvised. Moreover some of his ideas resonate strongly with more current ideas in cognitive science.

Jung viewed all mental experience as the product of four psychological functions: what he called sensation, intuition, feeling, and thinking. He grouped the functions into pairs, with sensations/intuition as one pair and thinking/feeling the other. He saw sensation and intuition as opposing ways of *apprehending* the external world, of taking in and integrating information. In contrast, he saw thinking and feeling as opposing ways of evaluating information.

According to Jung, although we each use all four mental functions, we each tend to rely on one more than the others. He classified people by the mental function they favor. The sensate perceiver is interested in the object of perception as it is in itself—in forming a more accurate sensory image of it as it is in the present, or in a usual or average state—in order to interact with it more effectively. The intuitive perceiver, in contrast, is more interested in mental apperception, or mental integration, than physical integration of the world. Although intuitive perceivers must, of course, form sensory images of the objects with which they interact to navigate physically, this is not their principle concern. Their mind is primarily interested in gaining information about an object or event to help them assimilate it mentally in all its uses, aspects, or meanings—to grasp it as a whole.

Each object or event is part of a pattern in space, because it is connected to other objects or events in the environment. It is also part of a pattern in time, because its connections to other things have changed in the past and will continue to change in the future. According to Jung, the intuitive perceiver is drawn to building a cumulative mental understanding of the patterns that objects and events make in space and time.[16]

These ever-enlarging patterns correspond to the *possibilities* associated with the object. The intuitive perceiver is interested in the sensory information or the data to the extent that it and the tacit knowledge associated with it help the imagination capture these possibilities. In contrast to the sensory perceiver, who tries to generate concrete representations or images that reflect the object as it is now, or in an average state, the intuitive perceiver tries to capture the complete profile of possible states of the object. For Barbara McClintock, for example, the coloration of each kernel of an ear of corn she harvested was part of a larger and increasingly complex pattern.[12]

Information about complex patterns of connections in time and space is taken in primarily by the unconscious, as Jung recognized. The conscious mind with its limited processing capacity is incapable of the large-scale associative, or parallel interactive processing required. As a consequence, what the intuitive perceiver takes in at an unconscious level may be much more important for the implicit goal of understanding the complete profile of possible states of whatever has captured their interest. This means that sharpening conscious attention does not always lead to more productive knowledge for the perceiver, and in some situations, it may lead to less.

The crucial knowledge for intuitive perceivers resides not only in the patterns of interconnection in the data but also in their interconnection with other patterns. This complex and largely unconscious web of information begins to map—to give shape to—what lies beyond. In contrast to the sensory perceiver, the intuitive perceiver is interested in the information largely as it points beyond itself to what is hidden.

The information the intuitive perceiver takes in provides fuel for imagination. However, this is not necessarily the unfettered imagination of an artist that can create new worlds. It may also be the more restrained imagination of a scientist that senses what is possible but still largely veiled from awareness. The movement beyond what is currently clear or known does not necessarily result in immediate conscious understanding. Yet as the information integrates below awareness, it often gains depth and salience for the conscious mind—in response to motivational/attentional forces that also function below consciousness. The growing knowledge may give rise to cognitive urges that express themselves as feelings, inarticulate cognitive shapes, or imaginative mental images.

## THE BRAIN'S TWO MEMORY AND LEARNING SYSTEMS

The brain has two different learning and memory systems, as McClelland and colleagues pointed out.[17] The distinction between them echoes aspects of Jung's distinction between sensory and intuitive perception. Both the hippocampal and neocortical systems are necessary for normal memory and cognition. They interact closely, although many the details of this interaction are still sketchy.

The hippocampal system encodes knowledge in a specific spatial and temporal context. It is responsible for episodic learning and memory, the specific what, when, and where of experience. The hippocampus is generally associated with declarative memory, or memory that can be stated. In addition, it is necessary to instruct, or to teach, many aspects of the neocortical memory and learning system, which generates and stores more abstract knowledge. The hippocampal system is also required to bring back into consciousness at least more recent memories from the neocortical system. Its neural encodings must be exceedingly plastic so they can change rapidly in response to recent experience. Likewise, hippocampal representations are necessarily narrowly tuned and distinct from each other because they generate concrete memories that can be recalled and stated.[18]

In contrast, neocortical representations are necessarily broadly tuned. They depend on more distributed or overlapping encodings, require repeated exposure, and change more slowly. Each encounter leads to widespread but incremental changes in connections. This more gradual formation allows the neocortex to abstract common elements between objects and between events.[19] The neocortex also generates much of the tacit knowledge that can be detected by implicit learning experiments. We know this because amnesiacs who cannot form declarative memories function as well as normal subjects on many tests of implicit learning.[20]

People who favor sensory perception likely depend somewhat more on the hippocampal *memory* system for their ordinary mental activity than intuitive perceivers. This is because of its grounding in concrete experience and its centrality to declarative knowledge. Likewise, sensory perceivers depend more on apparently conscious processing and synthesis.

In contrast, those who favor intuitive perception likely depend somewhat more on the neocortical memory system than sensory perceivers. The neocortical system gives rise to the patterns of change over time and space that Jung associates with intuitive perception. More of the intuitive perceiver's mental activity occurs below awareness than is true for sensory perceivers. What does make it to awareness is likely to be more abstract and apparently gist-like. Moreover intuitive perceivers may recall and process below awareness even material that could declared and processed consciously.

Because mental imagery corresponds to an inner seeing, the ordinary inner imagery of those who favor intuitive perception is also likely to be relatively enriched in neocortical content. It will be more abstract or and gist-like.[21] In contrast, the ordinary mental imagery of those who favor sensory knowing will be relatively enriched in hippocampal-dependent contents and thus more concrete in nature. Later in the chapter, I suggest a simple exercise that will allow readers to explore their own style of mental imagery.

## INTUITIVE VERSUS SENSORY SCIENCE

Scientific endeavor, as I have stressed, depends on both intuitive and sensory perceiving as well as imagery. However, at least at a superficial level, it is more aligned with sensory perception and imagery—with the data as it now stands and with models based on this data. In some scientific contexts, elaborating possible outcomes is viewed as mere speculation, favoring one over another is considered unwarranted bias, and drawing inferences based on such apparent bias may even be seen as dangerous. Yet these activities have an important role in the process of science, and all scientists engage in them at least some of the time. However those who favor intuitive perception are likely to do so more frequently because what is largely veiled and what goes deeper tends to be much more interesting to them than working with what is readily known.

When these intuitive assessments are backed by reason, they are more likely to be taken seriously by others. Yet those who favor intuitive perception cannot always report what makes them know that a certain outcome is likely, or that one explanation rather than another will prove right, especially on the spot. Their insight tends to depend to a greater or lesser extent on information that is not, or not yet, available to the conscious mind. Even when it is, as McClintock's struggles show, it is not always easy to get others to understand.

## POSSIBILITY VERSUS POTENTIALITIES

Jung recognized that intuitive perception depends on statistical patterns of spatial and/or temporal regularity recorded by the unconscious mind over time.[22] His view of intuition as seeing possibilities captured in a web of information that is primarily unconscious resonates with simple neural network theory and with simple artificial neural networks, like those that can mimic certain aspects of neorcortical leaning and memory. As the first part of Chapter 4 examined, the predictive ability of neural networks lies in their weighted patterns of connection, which reflect the past experience of the system as it develops over time. In the brain, these patterns of probability are encoded in the tendency, or disposition, of certain neurons to fire together. They allow for the pattern recognition essential to cognizing possibilities. Part of the power of Jung's understanding of intuition is that he grounded it in this statistically based associative knowing, which is contiguous with animal learning. Yet he also extended it to the highest reaches of human imagination, the mythic and spiritual realms with which his work is often associated.

Jung sometimes talks about intuition as sensing possibilities and sometimes as sensing potentialities.[16] The intuitive knower according to Jung seeks to know where something is going, in addition to where it has been. Yet this does not necessarily mean a simple projection of past patterns into the future. The idea of *potentiality* in contrast to *possibility* connotes change in the sense of development or unfolding, of something coming into being, or into awareness, that is latent or implicit at the current time.

Intuition as *capturing potentialities* is dramatically illustrated by the radical breakthroughs of scientific geniuses such as Archimedes and Einstein. It reflects the novel self-organization of perception that results from a substantial reordering of the cognitive landscape and leads to a large increase in cognitive capacity and meaning making. Rather than simple neural network theory, it is better explained by the more complex and dynamic model of brain function presented in the second part of Chapter 4, which can account for a profound reorganization of the cognitive landscape. Complexity theory can readily explain, for example, how all of the apparently unrelated individual possibilities associated with a lower level of understanding might be captured by the next higher level in a metapattern, or attractor. This metapattern encodes the full potentiality of the lower level and it may suddenly emerge in response to a small change. The cognitive reorganization might even occur in response to a small change in what the conscious mind would see as unrelated information. By the same token, this new level may begin to lay the groundwork for the next higher layer of potential knowledge.

Images can have an important role in revealing potentialities by incorporating unconscious knowledge and understanding and bringing them to awareness. Archimedes, so the story goes, under the threat of death had to determine whether the king's crown was solid gold. When he happened to see a baby placed in its bath (or in some versions, when he himself took a bath) he saw water spilling over the side. Realizing that water displacement provided a way to measure the volume of the king's crown, he ran out in the street (naked in some versions) shouting "Eureka!" Knowing the volume of the crown, along with its weight and the weight of gold, allowed him to determine whether the crown was indeed solid gold.

In whatever manner it may have occurred, the bath scene triggered the sort of sudden and radical reorganization of thought that often characterizes profound intuition. It allowed Archimedes to recognize that submerged (nonporous) objects displace an amount of water equal to their volume and to see its potential as a diagnostic tool. It also might also have laid the groundwork for his ideas of density and buoyancy.

Archimedes' "*Eureka!*" experience probably drew on a large amount of unconscious knowledge. He may have bathed in a tub himself as well as seen a number of children placed in their tubs. Even more important, all adults are likely to have an unconscious, if not semiconscious, sense that submerged objects tend to displace an amount of water equal to their volume and act accordingly. Barring occasional painful mistakes, we are careful not to place large amounts of anything in a full pot of scalding water. The sensory image of water spilling over the side when the baby was placed in its tub, or perhaps when Archimedes himself sat down in a bath, made the potential inherent in this implicit understanding with respect to his task salient and available to awareness.

Yet because this knowledge was implicit, and Archimedes had a pressing need for it, it could equally well have come to his awareness through a mental image rather than a current encounter. Indeed who is to say what really happened more than 1700 years ago, or if the story is even true? At a more basic level, it does not really matter how it occurred—if it did—because equivalent processes would have been involved. His need for a solution allowed him either to recognize the relevance of a current scene to the task of determining whether the king's crown was solid gold or to generate the scene in his mind, based on either a past experience or implicit knowledge.

Archimedes was likely a sensory perceiver, given the physical and practical nature of his interests. The concerns of the intuitive perceiver tend to be different, as a previous section stressed. Even when there is no dire need, their urge is to understand the object or event, not as it is in the present moment nor in some average state, nor to see a range of practical applications for it.

Rather it is to perceive the object or event in its full potential, in its wholeness. In moving toward understanding the whole of something, the intuitive mind largely below awareness seeks a new higher-level pattern that subsumes all possible individual patterns, implicit as well as explicit, associated with it. This metapattern may also connect with other apparently unrelated knowledge, either immediately or over time, and with it begin to suggest below awareness a yet higher level of understanding. The higher-level pattern may in turn recontextualize the lower-level patterns from which it is derived. Even if the potential novel understanding associated with the higher-level pattern does not come to awareness, intimations of it may sometimes be available to the receptive mind through mental imagery.

Full understanding of many things considered meaningful is likely to remain an elusive ideal. Nonetheless the intuitive perceiver tends to push forward. As a consequence, the quest of Jung and many others who favor intuitive perception frequently extends into the spiritual realm. Intuitive understanding of the spiritual realm can be transmitted to others only through imaginative images, such as the imagery or ritual associated with organized religion or the imagery that Blake and other mystics have used to communicate their experiences.

Sometimes the implicit goal of knowing the full potential of objects and events will orient a strongly intuitive scientist toward a holistic understanding of the cosmos. Each higher layer provides a new, larger context for the lower levels. The work of David Bohm, explored briefly in the next chapter, illustrates the thinking of a physicist with a strongly holistic perspective. He proposed a number of different layers of reality, with each higher level enfolding the lower ones.

The intuitive urge perhaps more often leads scientists in the opposite direction, toward an increasingly fundamental understanding of the natural world. The more the mind can perceive potentialities, the more likely it is to recognize that phenomena apparently sharing little on the surface are reflections of the same deeper pattern. Also the more the mind can perceive potentialities, the more it will be able to resolve apparently contradictory findings within a more penetrating level of understanding. The ability to perceive potentiality, for example, allowed Einstein to recognize the interconnection of time and space through the constancy of the speed of light in all frames of reference.

It is possible to intuit that two or more apparently similar situations are related or have a common meaning without being able to characterize their similarity. Understanding the underlying pattern may come only much later. This may depend on an imaginative image coming to mind, such as Einstein's teenage image of riding a light beam, which he himself acknowledged formed the germ of the special theory of relativity.[23]

## SPONTANEOUS MENTAL IMAGERY: LETTING THE MIND DO ITS OWN THING

Imaginative mental images are sometimes constructed at least partially with conscious effort. At other times, they may arise completely spontaneously. Spontaneous mental imagery tends to dominate awareness during daydreaming or when the mind is wandering.

Early empirical studies by Eric Klinger showed that people spend roughly a third of their waking time daydreaming or mind wandering.[24] Moreover he found that this apparently undedicated mental activity can serve an important function. Much of the time it consists of thinking about blocked goals and/or working out unresolved emotional issues.

During daydreaming, what is called the *brain's default network* is active. Certain brain areas that are activated during the resting or control state in functional magnetic resonance imaging experiments are *deactivated* during typical experimental tasks—tasks that involve attention to external stimuli.[25] These widespread regions tend to fluctuate in a coherent way at about 0.1 cycles per second and together are designated *the default network*.

Most of the brain's metabolic energy goes to support the default network and its functions. We know this because typical cognitive tasks imposed by experimenters only increase the brain's energy use by about 5%. The predominance of this "dark energy" in the brain is in accord with Freeman's work presented in Chapter 4, which suggests that most, if not all, of behavior and cognition are constructed from within. The default network also is activated in activities other than daydreaming, for example, thinking about the self.[26]

Approaching blocked goals during daydreaming or mind wandering depends on associative, or parallel interactive processing and its potential to combine things in novel ways. As Bar has pointed out, the default network overlaps considerably with the higher associative areas of the cortex.[27] It is activated during insight, which, like daydreaming, relies on associative processing.[28] Moreover, activity in the default network *before* subjects are given the three words for which they must find a common associate indicates whether they will use insight instead of a conscious approach to the solution.[29]

Brain imaging confirms that components of both the default and the executive network can coactivate when the mind wanders from an experimenter-determined external task,[30] as predicted by Klinger's finding that the daydreamer is often involved in solving blocked goals. The executive network is responsible for focused attention along with the cognitive control functions of the brain. Researchers are working out the many different potential interactions between the various components of the default and the executive network.[31]

As Klinger's work also hints, the distinction between a germinating intuition and daydreaming is not necessarily clear. The mental images that help to develop an intuition and carry it to awareness often appear aimless. Yet eventually they forge the relevant underlying strands of unconscious and unconscious information into a novel and cogent pattern.

Spontaneous mental imagery can also occur in response to a current conscious demand or need. If nothing is immediately forthcoming, we sometimes might deliberately access the less voluntary or default aspects of mind, as I did, for example, before the blink of the eye metaphor appeared as an answer to the question "What is intuition?" Purposefully turning toward the involuntary aspects of intelligence for insight requires a subtle alliance between the conscious and unconscious mind. As is often true with meditation, it may be necessary to use disciplined conscious focus to quiet the habitual activity of the mind.

## CAPTURING SPONTANEOUS MENTAL IMAGERY

Because internal mental imagery is such a basic part of the mental landscape, it may be hard to isolate. I propose a simple exercise that can help readers obtain a glimpse of their own favored style of mental imagery. It also illustrates the apparently contradictory demands on attention of intentionally turning toward involuntary mental activity.

The task is simply to let a spontaneous mental image of your grandmother come to mind. On your first several tries, you may find yourself directing your thoughts to produce a visual image of your grandmother. Capturing a spontaneous image may well require focused attention, but it should be deployed toward *receiving* whatever emerges from the mind. A number of trials might be necessary to get an image that feels like it just appeared in the mind.

Everyone's inner way of imaging their grandmother is likely to be different, and it may also vary from moment to moment. The style of imagery here is as important as its content. Which sensory modality is primary? Is it visual, and if so, is it weak or strong, scanty or complete? Is it static or is there motion to it? Does it include auditory elements?

My internal images frequently have visual elements, but they tend to be weak. Even with effort I am unable to bring a clear picture of my grandmother's face to mind. Instead, in an image that feels right, in front of a backdrop of heavy drapes, she is reaching out her arms holding something I can't see. I suspect she is holding a platter of the delicious garlic chicken she had ready for the mandatory Sunday visits of her eight children and their families. With reflection, I recognize that it is perhaps my most characteristic memory of my grandmother.

Even the image of my grandmother holding out her arms that I just mentioned involved considerable interpretation by my mind. In the picture that initially emerged, not only is the platter of chicken missing, but her arms and the drapes are as well. Instead I see only a pared-down sketch suggesting an action: just two straight lines extending diagonally forward with some heavier vertical lines behind.

I strongly favor intuitive perception. This sparse style of visual imagery is consistent with the view that those who favor intuitive perception tend to have more abstract or gist—like imagery than sensory perceivers, at least with regard to their ordinary mental activity. According to Jung, the least active function is the opposite of the one used most.[15] The sensory function of those who, like me, strongly favor intuitive perception will have considerably less influence on their usual perceptual experience and thus their usual mental imagery. Conversely the intuitive function of strongly sensate people will have considerably less influence on their ordinary experience.

Yet intuition is not limited to intuitive perceivers, as the *"Eureka!"* story suggests. Moreover the preceding discussion does not necessarily hold for mental images that bring intuitions to awareness. Although the sensate function of intuitive perceivers does not have a lot of influence on their ordinary cognition, it is still active in their unconscious. In moments of intuition when the unconscious is more or less unopposed, the least used and least developed mental faculty (the opposite of the favored one) may break into awareness. However it tends to take a different form than in those who favor that faculty for ordinary cognition. It also can bring with it a sense of freshness, spontaneity, and vitality.

When I tried again to catch a spontaneous mental image of my grandmother, I was rewarded with an intuitive image that told me something important I had not realized before. It was also more concrete and sensory than my usual mental imagery—albeit in its own special way. I saw a strong, concentrated pulsation on the upper left side of what must be her face. The image includes explicit movement but it evokes "energetic" rather than spatial movement. My sense of its authenticity was strengthened because I found it so compelling and enigmatic. Eventually I decided it pointed to her human force, or even better, her "canniness," reflected in part by her strong hold over her eight children. Those who favor sensory perception and concrete imagery in their ordinary mental activity may also find that their intuitions come to awareness in an altered style of mental imagery.

## ACTIVE IMAGINATION

Jung the clinician championed what he called *active imagination*.[32] He asked patients to elaborate on a dream image, or one of their drawings, without censoring or trying to make sense of what they said—in effect asking them to enter into an extended *blink of the eye/I*. This imaginative train of thought was intended to help them discover the underlying unconscious dynamics of their situation, *the complex* or chaotic attractor around which their emotional life most resolved.[33] The sequence of imaginative images might also reveal clues to the unfolding of a solution that could help them reintegrate their past experience and regain more of their full potential. According to Jung, these images come from what he and others have called *the Self*, an aspect of the intuitive self that strives to integrate the conscious and unconscious aspects of mind.

As we attempt to understand a novel phenomenon or problem, our unconscious mind sometimes brings to awareness an imaginative image that can reveal unexpected potentialities. Images referring to the body tend to be particularly powerful. Einstein's adolescent notion of riding on a light beam, for example, or the different body-based images my unconscious gave to the question "What is intuition?" use the body as a screen upon which to project novel ways of thinking about a problem. The next chapter explores the importance of body-based imagery in intuition in greater detail.

An organization called *Synectics*, founded in the 1950s by William Gordon and George Prince, championed the idea that a more direct effort to harness the unconscious mind would enhance scientific creativity.[34] Synectics encouraged the use of imaginative imagery, especially body-based or kinesthetic imagery, to solve scientific or technical problems. I have summarized a lengthy transcript of an inventor using the technique to come up with the internal mechanism for a display dial on an airplane altimeter. Hired by the government, his assignment was to design a horizontal display without the friction generally associated with horizontal indicators.[34(pp16–18)]

While the inventor is taking apart an old altimeter, he notices that everything but its spring seems to disappear from his mind. He then wonders what it means to be a spring. Taking the problem into his body, he asks, if he were a spring how would he feel? He toys with the idea of folding in and expanding out in several different guises.

The inventor then finds himself thinking directly about the answer to the problem. He warns himself that if he expects to get an anywhere, he has to concentrate on the spring, not the eventual solution. He has to remind himself of this a number of times during the session.

Again and again, he plays with the notion of a spring and springiness. For example, he associates to the seasons and how they spring in and out. Eventually he recognizes that the motion of a spring winding in and out will be essential. A bit later he sees that the motion of the spring has to do everything, and then he realizes that he needs to get an arc out of the spring's motion.

In an effort to get inside the spring's motion, he takes the problem into his body again. This time he imagines being attached to a spring as big as a house and holding on to its free end as it contracts and expands. He then sees that if he put a blob of ink on the end of the spring as it moves, it will provide him with the just the arc he needs.

Just as he comes to the solution, he marvels that the idea no longer feels like his creation, or part of him. It has taken on a life of its own. He wonders if this is what authors feel when they say that their characters write their own parts.

The inventor's body and imagination provide many different images of springiness throughout. Especially with the full transcript, the logical mind may feel impatient with the redundancy and conclude that the inventor is not getting anywhere. Yet an alternative, nonlinear perspective suggests that the underlying core of the insight is gaining force.

Intuition is often viewed as cutting through to the heart of the matter. The final knowing does go to the core, and it sometimes appears in a flash. However it is often preceded by a period in which a variety of loosely related images and ideas emerge. Jung called this meandering process *circumambulation* to suggest that one circles around to encompass the whole.[35]

Intuitive images often work to reveal potentialities by imparting meanings that appear logically disjunctive to an event. This, in turn, can force the mind to go deeper to find the connecting links. The inventor, for example, associates the *springiness* of the mechanical *spring* with the seasons and the way they go in and out. Even when imaginative images that appear are opaque to the conscious mind, they may still inform understanding below awareness.

Because of the apparent openness and lack of rigor of imaginative images on the one hand and their potential hidden unconscious complexity on the other, they often lead back and forth between the unconscious and conscious mind. The insight, or the next layer of insight, underlying an image might appear only after a while. Such slow-release images can act as a long-term guide, with layer upon layer unfolding over time. An especially fertile intuitive image such as "a blink of the eye/I" can forge a semipermanent channel between the two aspects of mind. Because it keeps pointing beyond itself to the deeper and more complete patterns it represents, it functions as a nonlinear road map on the intuitive journey.

Scientists welcome intuitive insight when it comes in analogies that feel right to the conscious mind. However they largely shun the use of spontaneous fanciful imagery as an aid in their thinking, considering it more suited to artistic or mystical pursuits than the rigorous task of investigating the natural world. Also, yielding fully to the potential richness of intuitive imagery, as the inventor emphasizes, means letting go of the direct quest for the solution for the interim and entering a different mental realm. This is not necessarily easy for anyone, and perhaps especially for the scientific mind.

Yet images that seem to defy reason are not necessarily at odds with the rational agenda of science, because they *augment* rather than *supplant* rational thought. *Knowing* is something different from having even multiple levels of mental imagery come to consciousness. An infinite number of different specific images can all point toward the same insight. Each is either a realistic or a fanciful particular expression of the intuitive knowledge, which itself subsumes them all. They are all just pointers to that moment of convergence we call knowing.

My intent is not to convince anyone to employ active imagination as a tool. It is rather to interest readers in attending to the subtle promptings of their intuition. Knowledge of active imagination is useful because it highlights how intuitive imagery emerges from the unconscious. Also it provides a glimpse of the deep purposefulness that can drive intuitions. No matter how peculiar the images that appear may seem to the conscious mind, they have logic of their own and tend to lead toward solutions. Intuition is a goal-directed, or intentional, process that serves an organism. To the extent that our conscious mind can relinquish control and allow unconscious images to arise, we can experience the force of intuition in a relatively pure fashion.

With the experience of intuition as a force, we are more likely to become attuned to its subtle promptings in the midst of ordinary functioning. It is then possible to think of the mind in a new way—as a potential conduit for the intuitive force. We register its messages earlier, trust them more, and even listen for them. When we respect the subtle images or feelings that emerge unbidden from the unconscious mind as much as we do our reason, we are likely to experience our understanding coming through us. In these moments, we do not simply use our mind, but honor it as well.

## REFERENCES


1. Strathern P. *Mendeleyev's dream: the quest for the elements*. New York: Thomas Dunne; 2001.
2. Hadamard J. *The Mathematician's mind: the psychology of invention in the mathematical field*. Princeton, NJ: Princeton University Press; 1973.
3. Feynman R. *QED: the strange theory of light and matter*. Princeton, NJ: Princeton University Press; 1995.
4. Keiser D. Physics and Feynman diagrams. *Am Sci* 2005;**93**:156–65.
5. Morris JS, Dolan RJ. Involvement of human amygdala and orbitofrontal cortex in hunger-enhanced memory for food stimuli. *J Neurosci* 2001;**21**:5304–10.
6. Freeman WJ. *How brains make up their minds*. London: Weiden & Nicolson; 1999.
7. Wheeler ME, Peterson RE, Buckner RL. Memory's echo: vivid remembering reactivates sensory-specific cortex. *Proc Natl Acad Sci USA* 2000;**97**:11125–9.
8. Palmiero M, Belardinelli MO, Nardo D, Sestiere C, Di Matteo R, et al. Mental imagery generation in different modalities activates sensory-motor awareness. *Cogn Process* 2009;**10**(Suppl. 2):268–71.
9. The American Heritage Dictionary, Boston, 1991, Houghton Mifflin Company, 1981.
10. Damasio A. *Descartes' error: emotion, reason and the human brain*. New York: G. P. Putnam's Sons; 1994.
11. Bargh JA, Chaiken S, Raymond P, Hynes C. The automatic evaluation effect: unconditional automatic attitude activation with a pronunciation task. *J Exp Psychol* 2005;**32**: 104–28.
12. Keller E. *A feeling for the organism*. New York: W. H. Freeman; 1983. p. 126.
13. Kekulé FA, reprinted in Japp FR. Kekulé memorial lecture. *J Chem Soc* 1989;**73**:97–138.
14. Popper K. *The logic of scientific discovery*. New York: Basic Books; 1959.
15. Jung CG. *Psychological types: (collected works vol. 6)*. [Hull R, Trans.]. Princeton, NJ: Princeton University Press; 1971.
16. Jung CG. Psychological factors determining human behavior. [Hull R Trans.]. In: Jug CG, editor. *The structure and dynamics of the psyche, (Collected Works vol. 8)*Princeton, NJ: Princeton University Press; 1971. p. 122–3.
17. McClelland JL, O'Reilly RC, McNaughton BL. Why there are complementary learning systems in the hippocampus and neocortex: insights from the successes and failure of connectionist models of learning and memory. *Psychol Rev* 1995;**102**:3419–57.

18. For a study showing that the hippocampus might also be involved in some implicit mental activity see Greene A. Human hippocampal-dependent tasks: is awareness necessary or sufficient? *Hippocampus* 2007;**17**:429–33.
19. Rumelhart DE, McClelland JL. Parallel distributed processing: explorations in the microstrucutre of cognition. *Foundations*, vol. 1. Cambridge, MA: MIT Press; 1986.
20. Willingham DB, Neissen MJ, Bullemer P. On the development of procedural knowledge. *J Exp Psychol Learn Mem Cog* 1989;**14**:1047–60.
21. For a study showing that the cortical system can give rise to mental imagery even in the absence hippocampal participation see Stickgold R, Malia A, Maguire D, Roddenberry D, O'Connor MC. Replaying the game: hypnagogic images in normals and amnesiacs. *Science* 2000;**290**:350–3.
22. Most of Jung's writing on intuition predated Donald Hebb's work, which linked associative learning to the plasticity of neurons and was central to early work on neural networks. Hebb D. *Organization of behavior*. New York: John Wiley; 1849.
23. Holton G. *Einstein, history and other passions*. Reading, MA: Addison-Wesley; 1996. p. 199.
24. Klinker E. Daydreaming and fantasizing: thought flow and motivation. In: Markman KD, Klein WMP, Suhr JA, editors. *Handbook of imagination and mental simulation*. New York: Psychology Press; 2008. p. 225–50.
25. Raichle ME. Two views of brain function. *Trends Cogn Sci* 2010;**14**:180–90. The default mode includes the medial prefrontal cortex, some of the medial temporal lobe, the precuneus, areas of the parietal lobe and the posterior cingulate cortex.
26. Andrews-Hanna JR. The brain's default network and its adaptive role in internal mentation. *Neurosci* 2011;**18**(3):251–70. https://doi.org/10.1177/1073858411403316.
27. Bar M, Aminoff E, Mason M, Fenske M. The units of thought. *Hippocampus* 2007;**17**:420–8.
28. Jung-Beeman M, Bowden EM, Haberman JL, Frymiare J, Aerameel-Liu S, et al. Neural activity when people solve verbal problems with insight. *PLoS Biol* 2004;**2**:0500–10.
29. Kounios J, Frymiare JL, Bowden EM, Fleck JI, Subramaniam K, Parrish TB, et al. The prepared mind: neural activity prior to problem presentation predicts subsequent solution by sudden insight. *Psychol Sci* 2006;**17**:882–90.
30. Christoff K, Gordon A, Smallwood JR, Smith R, Schooler JW. Experience sampling during fMRI reveals default network and executive system contributions to mind wandering. *Proc Natl Acad Sci USA* 2009;**106**:8719–24.
31. Christoff K, Irving ZC, Fox KCR, Spreng RN, Andrews-Hanna JR. Mind-wandering as spontaneous thought: a dynamic framework. *Nat Rev Neurosci* 2016;**17**:718–31.
32. Jung CG. In: Chodorow J, editor. *Jung on active imagination*. Princeton, NJ: Princeton University Press; 1997.
33. The complexes represent restricted patterns that tend to limit behavior to one aspect of the archetype, which reflects an even higher order pattern. The goal is to free behavior so a person can experience all the different aspects, or the full potential, of the archetype. Both complexes and archetypes can be seen as chaotic attractors.
34. Gordon WW. *Synectics*. New York: Harper & Brothers; 1961. p. 16–7.
35. Jung CG. *Analytical psychology: its theory and practice: the tavistock lectures*. London: Routledge; 2014.

# CHAPTER 7

# The Importance of Embodied Experience and Imagery in Intuition

## Contents



Subjective experience has an important place in intuition. As the last chapter emphasized, intuition depends on mental imagery that is often suffused with subjectivity. This chapter probes more deeply into the first-person grounding of intuition, using another intuitive experience I had.

The "blink of the eye" encounter intimated that subtle internal physical experience might have a pivotal role in intuition. Other hints also pointed to the special importance of body-based imagery in intuition. I pondered why this might be so for over a year before I had an experience that suggested an answer.

## A SENIOR MOMENT

After a visit with friends in France, I went to explore Provence for a few days on my own. Late one morning, I set off by car to visit the extraordinary city of Les Baux, an ancient town partly hewn out of rock. Sharp outcroppings of gray bauxite gradually came to dominate the terrain on the long, upward-climbing, curvy road, eventually fusing into a phantasmagorical moonscape.

*Understanding Intuition*
ISBN 978-0-12-814108-3
https://doi.org/10.1016/B978-0-12-814108-3.00007-5

Then all of a sudden, there it was—a stark, craggy rock citadel like nothing I had ever seen!

The impressive, steeply rising, massive rock city towering high above the cars and buses in the parking lot had an impenetrable air. Instead of eating lunch there as I had planned, I opted for a leisurely meal in a restaurant suggested by my guidebook. It was housed in an old mansion, surrounded by a lush garden, halfway down a canyon on the far side of the citadel. I had a book with me and thought I might have a good read before breaching the gigantic rock bastion.

My interest in the role of metaphor in intuition had led me to a book by George Lakoff, a linguist at University of California at Berkeley, called *Women and Fire and Dangerous Things*.[1] It explores category formation and the role of metaphor in both language and conceptual development. The unusual title refers to some of the items in the category *Balan*, one of only four categories in an Australian Aboriginal language. Balan also includes the sun, the stars, and anything connected with water.

A number of years ago, Lakoff, along with philosopher Mark Johnson,[2] proposed that language and conceptual development both result from the metaphorical projection of basic *kinesthetic schemas* into the abstract realm. Kinesthetic schemas are open-ended relational patterns, or gestalts, that reflect the basic organization of the body as well as its potential for interaction. In conjunction with the various possible modes of movement of the body, they give meaning to our interface with the environment. They also help us carve the world into natural categories, or "kinds," based on similarity of interaction (Chapter 4).[3] They come about because we are separate objects in space with a specific body plan and ways of moving. The common denominators of all experience, the kinesthetic schema are functional in the newborn or require only minimal developmental experience. They also reflect social interactions because an infant's development depends on its connection with at least one other person.

In *Women and Fire and Dangerous Things*, Lakoff emphasizes five of the basic kinesthetic or embodied schemas and demonstrates how rational mental activity could have developed from them. They include the *container schema*, the *part–whole schema*, the *link schema*, the *center–periphery schema*, and the *source–path–goal schema*. Other higher animals have similar embodied patterns but put them to more restricted use at least in the higher cognitive realm. Lakoff does not argue against the existence of a separate rational mental function; rather he shows that it is unnecessary to account for human rationality.

Each general pattern of embodied experience has its own inherent parts and organization. They are evident in the names of some, such as the *center–periphery* or *part–whole schema*. In others, such as the *container* or the *link schema*, they are only implied. The container schema, for instance, implies an interior, an exterior, and a boundary. In the womb we experience ourselves as being contained, and after birth we begin to experience ourselves as a container, no doubt one sometimes filled with difficult-to-manage needs and impulses. We come to know the kinesthetic schemas not by thinking, but from bodily experience—from the inside.

These embodied patterns also determine our perception of spatial relationships in the outer world. For example, the container schema gives rise to the notion of *in* and *out*. Eventually, according to Johnson and Lakoff, these and other preconceptual schemas govern developing conceptual structure, and thus how we think about the abstract realm. The basic principles inherent in our embodiment come to organize both own mental space and our understanding of the external world. As a linguist, Lakoff seeks evidence for this view in the metaphorical projection of these and other kinesthetic schema into language. For example, he traces most of the idioms that use either *in* or *out*, such as "walking *in* a daze" or "setting *out* the dishes," to the container schema.

This is abstract and heady stuff! At the time I couldn't find my bearings in the material at a conscious level, although my absorption and excitement suggested I was resonating with it strongly below awareness. The last thing I remember thinking was that I didn't know what was meant by *conceptual structure*. Granted conceptual activity must have a structure, but my mind knew no way to look beyond its own current activity to appreciate this structure, especially in any overarching sense. Yet just recognizing this difficulty felt like a significant breakthrough in my ability to relate to this material on a conscious level.

As I looked away from the book to reflect more on this intriguing obstacle to conscious understanding and verification of its thesis, my eyes must have wandered to the ancient stone walls of the dining room. When I returned to self-awareness, I found myself fixated on the arched stone doorway connecting the dining room to the large entrance hall. I was fascinated by how the small rectangular stones at the sides of the arch gradually gave way to wedge-shaped ones as the arch began to curve.

My gaze took in more of the walls of the dining room and also the hallway with its stone barrel roof. I observed them with transformed eyes, as an idea apparently planted during my senior moment took root. Stone upon stone, just a few different shapes in careful combination can make an arch. Several arches each opening up to the next can support a barrel roof. Arches in increasingly sophisticated combinations give rise to the glories of Romanesque and Gothic architecture. How could I have missed the elegant simplicity and power inherent in the construction of a stone arch before? I wondered.

I had never thought much about construction principles, and certainly not about the way ancient stone buildings were built. I now recognize that awareness of building principles along with construction skills reflects the activity of what Jung called the *sensate mental function.*[4] According to Jung, as I mentioned in Chapter 6, a strong intuitive mental function, which is grounded in a proclivity for implicit or unconscious learning and inference, is generally paired with a correspondingly weak grasp of sensate matters. The mental faculty that is the least dominant at a conscious level instead influences mental activity from below awareness. When it does break through to consciousness, it tends to express itself in a different way than in those who usually favor the faculty. It often brings a sense of freshness, exhilaration, and empowerment.

Indeed my insight at Les Baux gave me the feeling that everything in the past that appeared vague might now be within my grasp; I saw the whole world anew. At a much-rebuilt church at the top of the citadel, I confirmed what I now realized must be true. Columns, too, are built stone by stone, sometimes like the ones holding up the portico of the church, from one small round slab upon another. A few days later, when revisiting Notre-Dame and other stone marvels in Paris, I appreciated them with different eyes. They seemed even more magnificent when viewed through the conscious lens that they had been built stone by stone, based *ultimately* on a few elementary construction principles.

## Drawing a Blank

Only months later did I realize that my epiphany about ancient stone building techniques at lunch that day probably had some connection to the Lakoff and Johnson material I had been reading immediately before, and also to being at Les Baux. Although I felt the pressure of these connections just below the surface of my awareness, I could not capture them. A friend with whom I shared the story, a linguistics professor well versed in the work of Lakoff and Johnson, felt the same pressure and also could not capture the underlying connections at a conscious level.

During this same period, I was struggling with another question as well. Why can body-related movement sometimes be used apparently interchangeably with the movement of external objects in intuitions? In the "blink of the eye" encounter, for instance, the subtle sensation of a constriction and then expansion in my body was accompanied by a mental image of a signal first going *down* and then *up* on a graph or an oscilloscope. Eventually I realized that *Women and Fire and Dangerous Things* spoke to this curious "slippage" between internal and external experience, and I began to reread the book.

This time I understood the material at a more conscious level, although at first I still found it too far from my direct experience to integrate with my own thinking. But when I remembered my extraordinary experience at Les Baux, everything began to fall into place. The connections between the Lakoff–Johnson material and my epiphany about early stone construction then appeared so obvious that the fact that neither my friend nor I had been able to capture it seemed an important part of the story itself.

## Inside Out

With my experience at Les Baux, my unconscious had provided an extremely powerful experiential metaphor or analogy for the Lakoff and Johnson material. As is the wont of unconscious images, it offered a number of different levels of overlapping insight. At one level, the experience provided a physically grounded metaphor for the material I was reading at lunch that day and found too abstract to comprehend consciously. *Conceptual structure* is like *building structure*. Both are essential to structuring *spaces* (one mental spaces and the other physical spaces) and both depend on the same basic relational patterns or schemas. Both can also combine these relational schemas in various ways to make an infinite number of different structures.

The metaphor of physical structure, by reifying the abstract idea of conceptual structure, a concept I had never considered before, freed my mind from the seemingly boundless task of thinking about itself and its potential products. Substituting something concrete and external that could be encompassed more easily allowed my mind to get out of its own way. Of course, I did not *get* this extraordinary metaphor at the time. Yet none of its power was lost during the delay. I suspect that puzzling over these issues allowed me to appreciate more fully the answers that finally came. This may be true more generally, as some experimental evidence suggests.[5]

My attention was caught by an arch, an important form of the link schema. Arches can function simultaneously at a number of different levels in a structure. For example, they link separated horizontal spaces and they also provide a vertical weight-bearing link that can support the upper parts of a structure. In a similar way, each of the kinesthetic or embodied schemas can have a number of simultaneous and interwoven roles in conceptual structure.

In a sense, the link schema is the core schema for all structure, physical or conceptual. It is inherent in the very idea of connection, or relationship, and therefore it is at the base of the other schemas. Just as stones linked in specific relationships to each other allow the construction of potentially durable and elaborate physical structures, mental links in specific relationships to each other allow the construction of potentially durable and elaborate mental structures. Rational conceptual structure depends on one or a series of links that proceed from premise to conclusion in an orderly way. Intuitive conceptual structure, in contrast, depends on a number of different, or distributed, but interwoven links that together form a whole. Like intersecting arches, intuition provides many points of grounding for the multiple simultaneous interactions that come together to shape the whole.

The metaphor was perhaps even more powerful at a process level. By redirecting my attention to the physical world, my senior moment at Les Baux in part illustrated the larger point Lakoff's book made, and by the same token eventually helped to verify it intuitively. By switching my attention to something as basic to civilization as the development of stone masonry and architecture, it mirrored the unconscious redirection from inner experience to the outer world that happens as we humans explore and exploit our environment. This redirection also allows us without being aware of what we are doing to project our own bodily experience to structure our understanding of the external world and also to structure what we create in the world. What we see therefore mirrors knowledge derived from our own embodied experience back to us. This exchange is not only "legitimate," it is also an important source of our power as a species.

Other primates can do this to a more limited extent. Yet just as early humans did, some animals seek shelter from the elements in caves. With its unusual cavernous architecture, Les Baux was an integral part of this unusual experience because it resonates with our evolutionary past as animals. It also evokes the transition when we began to be able to exteriorize our embodied knowledge and "build" upward both physically and conceptually. By the same token, it represents the impenetrability of the preconceptual realm where this transformation still continually occurs.

The problem is that for the most part we remain unaware that any redirection has occurred and instead experience a reality that is out there and independent of our own inner experience. My inability as well as my friend's inability to capture the connection between the Lakoff material and my experience that day at Les Baux illustrates the difficulty we have in making the linkage between our inner and outer world. Bodily sensations, kinesthetic schema, self-awareness, and even thoughts about conceptual structure are subjective in the sense that they condition our experience from the inside. In contrast, stones, building principles, and the structures they result in are objective in the sense that they are about the external world and its properties. The connection between the two realms is not lost to us. However because of our commitment to the separation of the objective world from the world of inner experience, the connection tends to remain shrouded in the unconscious. We may sometimes recognize it as a slight pressure or very faint echo of familiarity in the background of experience from which the objective world arises.

My "blink of the eye" encounter, which occurred in response to the question, "What is intuition?" underlines the importance of this shadowy slippage between the inner and outer realms to understanding intuition. The apparently divergent images that appeared from my unconscious mind when I looked inside all shared a common pattern of motion: first a restriction or closing, *a down*, followed by a reversal or an opening, *an up*. It was this felt similarity of motion that made me recognize the coherence of the different images and thus their validity as an answer to the question, even as I did not understand what any of them might mean. As Lakoff points out, metaphorically speaking, *up* generally means *more* and *down less*.[1] The down–up pattern is useful for thinking about a number of different aspects of intuition. Yet at a process level, the experience seems to be saying, at least in part, that intuition about difficult questions can come to awareness in multiple different images that encode the same abstract patterns of change in both embodied experience and the movement of externals. These dynamic patterns grounded in our bodily experience have a core role in intuition because they relate us to the external world at the same time that they guide and unify cognitive experience. The different images that can encode an intuition together with the deeper patterns that unite them can greatly enrich as well as transform our usual understanding.

## THE DEVELOPMENT AND ONGOING INTEGRATION OF BODILY EXPERIENCE AND COGNITION

We are built for action—for intentional movement. It is part of our evolutionary heritage as animals. Our considerable intelligence depended on the development and augmentation of the frontal lobes of the brain, which evolved from the motor cortex. The frontal lobes confer the capacity to shape action through anticipation and planning. All cognition can be viewed as an adaptation to support more successful inter-*action* with the environment, which includes other people, to satisfy survival and other needs.

Motor-related activity and cognition continue to integrate with each other throughout life. Observing man-made objects such as tools, seeing pictures of them, imagining them, or silently reading their name all activate the premotor region of the frontal lobe, which is crucial to the control of movement.[6] Moreover the specific areas activated are the same ones that are activated during the actual manipulation of these objects. Substantial evidence also suggests that language comprehension about action uses some of the same neural circuits as the action itself.[7,8] For example, subjects reading a phrase that implies outward motion, such as "open the drawer," comprehend it more quickly when they are required to make a motion away from their body, a concordant motion, as opposed to one toward their body, a discordant motion.[9] Sensorimotor circuits are also activated during the creation of novel metaphors that extends action into the abstract realm.[10,11]

When we imagine or think about action, our brain seems to simulate it. Some researchers also believe that to perceive purposeful behavior in others, we have to simulate it below awareness in our own brains.[11] This view rests in large part on the neurophysiology of groups of neurons in the premotor region and the parietal lobe that map the body's relationship to objects in space.[10,11] (With simulated action, the connections between the actual motor neurons that would carry it out are thought to be inhibited.)

Take *grasping*, a basic form of intentional behavior that is present in infants. Gallese and others have shown that certain cells in the monkey's premotor cortex fire whenever the animal successfully grasps something, such as a cup or a slice of apple, whether by mouth or with one or two hands.[12,13] Another group of grasping-responsive cells in the premotor region has visual properties in addition to motor properties. Called *mirror neurons*, they are active whenever the monkey grasps an object or sees the experimenter grasp one. They also fire when the monkey or the experimenter grasps something the monkey saw placed out of sight behind a barrier. If a habitual action involves a sound (for example, cracking a nut or tearing a piece of paper), special cells in the premotor region called *audiovisual motor cells* respond to the sound as well as to the sight and motion of the action. Another group of premotor neurons with both visual and motor properties fire in response to the presence of any graspable object with a particular shape and size, and/or kind of grip, even when neither the monkey nor the experimenter reaches for it. Moreover, a group of cells in the parietal lobe fires when the monkey turns its head a certain amount (say, 15 degrees to the right) or when an object is placed in that position.

Together these grasping-responsive and ancillary cells encode the monkey's multimodal sensorimotor generalizations—in other words, their meaning making—around grasping. They communicate in a *nonlinguistic manner* that either the monkey or someone else in their immediate vicinity, grasps or can grasp an object with a certain particular grip and size and shape, in a certain position relative to its head. Various other groups of cells in the premotor and parietal areas encode the many additional details required for successful action, such as the specifics of the motion and the force required for the beginning, middle, and end stages of grasping. Bound into networks, the various cell types crucial to organizing complex grasping behaviors in the monkey through stimulation also likely have an essential role in perceiving grasping behaviors and making inferences about them in others.

The monkey mirror neuron studies reveal the important point that the close equivalence between internal and external meaning making, at least in monkeys, is hardwired in the brain. By the same token, the multisensory, grasping-related premotor and parietal cells indicate that the monkey brain is genetically prepared to generalize from experience and encode critical abstract regularities. Referring specifically to the audiovisual motor neurons, but applicable to many of the sensorimotor neurons, Gallese writes:

> *[T]hey seem to suggest that it is possible to have the sameness of informational content at a quite "abstract" level, the level of conceptual content, without being endowed with the cognitive faculty of language.*[15(p1238)]

Evidence also points to the existence mirror neurons in humans, but there is limited opportunity to record from single cells in the human brain.[10] Some argue that with humans as opposed to monkeys, associative learning mechanisms rather than mirror neurons forge the connections between cells in the sensory and motor areas that respond to one's own action and to similar action by others.[16] In either case, our brains build a model of the external world that integrates our sensorimotor experience. Because of its adaptive value, the equivalence between inside and outside that supports concept formation is, in one way or another, part of our evolutionary heritage.

As Gallese goes on to say, the growth of conceptual knowledge occurs automatically because of the dynamic nature of the brain's model of the environment. It is continually reshaped by interaction with the world. Even in the absence of language, the model increasingly incorporates abstract knowledge that supports the implicit inferences so central to intuition:

> *[T]he level of conceptual knowledge that I am proposing to ascribe to monkeys (but that I also take to be quite alive in our human mind), thus enabling them with the possibility to entertain certain abstract contents, is heavily dependent on "implicit inferences." Inferences are just more or less reliable predictions about fact. And prediction is a product of the constant shaping/rewiring of our model of the world as we interact with it.*[15(p1238)]

Human nonverbal conceptual abilities, as Gallese intimates, are much greater than those of monkeys. Even before our ancestors developed language, we were already a highly intelligent and innovative evolutionary line.[17] A portion of our behavioral insights still may not require verbal abilities. The embodied mind has a reservoir of knowledge that is not directly available to consciousness. Intuition works in our attempts to understand "objective" reality, in part because it allows us to exploit and build on the body's inherent nonverbal understanding of the world in which it exists.

Unlike monkeys, we do have language, which greatly magnifies our conceptual abilities. Moreover we can manipulate language in both associative and rigorous ways. The rich multidimensional knowledge that comes from our experience in the world, in conjunction with our internal mental activity, sometimes can be grasped and expressed only via metaphors, which give voice to these intuitions.

For both humans and monkeys, sensorimotor concepts such as grasping embed complex combinations of the more basic kinesthetic schema highlighted by Lakoff—some of the elementary components of our interaction with our environment. Survival requires that the newborn grasp the milk-giving breast or nipple. As intentional activity, it represents the most basic form of the source–path–goal schema. To grasp something also means to form a link with it, which is essential for the infant at many different levels. Grasping the breast also involves the container schema, because it forms a link with something outside and brings it inside. In humans, via metaphorical extension this sensorimotor knowledge eventually supports language formation and reasoning. Then we use verbal metaphors to grasp for, in the sense of reaching for, abstractions we sense below awareness but cannot express directly.

## A Two-Way Street

Most cognitive scientists are interested in the development of higher cognitive abilities, and therefore they have tended to stress the movement of influence and understanding via metaphor from the concrete sensorimotor to the abstract conceptual realms. Yet my experience at Les Baux highlights that more abstract levels of cognition (for example, the notion of *conceptual structure*) can conversely via metaphor influence much more concrete levels of perceptual attention, experience, and understanding. This is now supported by a number of reports in the cognitive science literature. One of the earliest showed that subjects asked to recall an unethical act were more likely afterward to choose antiseptic wipes over pencils than subjects who were asked to recall an ethical act.[18] Another showed that subjects primed to feel social isolation judged the ambient temperature to be lower than did control subjects.[19]

Social isolation and morality are highly charged emotional concepts. Emotions strongly influence perception; thus these effects in part hitch a ride on the emotional system, which is an inherently biasing mechanism. However at Les Baux I was grappling with the notion of conceptual structure, which connotes abstraction in and of itself and is emotionally neutral. Nonetheless, via unconscious metaphorical redirection, the next thing I knew, I found myself fixated on the arched doorway of the dining room, appreciating its construction. This suggests that even without a substantial boost from emotion, conceptual experience through metaphor can influence concrete perceptual experience.[20]

The connection between abstract conceptual understanding and concrete perceptual experience via metaphor even without an overt emotional link is bidirectional, or a two-way street. Our inherent action orientation sometimes works in conjunction with the inverse of the Lakoff and Johnson theory discussed before. Just as we project concrete experience into the abstract conceptual realm via metaphor, we can also reverse the process and experience the abstract conceptual realm through concrete perceptual experience and imagery.

As Gallese writes, "the representational dynamic of the brain can be conceived of as a type of self-organization, in which action plays a crucial role"[15(p1231)] Bidirectional connections between higher and lower levels are central to self-organizing systems, as the second half of Chapter 4 emphasizes.

It is not so surprising then that concrete perceptual experience can sometimes function as readout for abstract knowledge. Moreover the reverse pathway can begin with abstractions that would seem to transcend any concrete perceptual experience even via metaphorical extension by language, such as Einstein's theory of special relativity. This also means that insight about what might seem daunting abstract questions can hide in the strange twists and turns of one's attention to the world, as my experience at Les Baux illustrated.

## Two Modes of Intuition: Inside Out and Outside In

What has been said so far suggests that most knowledge is *ultimately* grounded in intuition: in sophisticated but hidden meaning-making processes. At the same time, it evokes two different types of intuition, although only one is generally labeled as intuition. Both are encoded in the phrase "a blink of the eye/I." In the *inside-out* mode, we project bodily patterns outward via metaphor when trying to understand something new about the external world. We then tend to perceive these patterns as being objective and independent of our inner experience. The *blink* associated with this sort of intuition is an instant of blindness as well as creativity that transforms the *I* into *the eye*. To a greater or lesser extent, it is a component of *the eye-space*, or sensory mode of perception often associated with the objectivity of science.

The work of Lakoff and Johnson (and in part my experience at Les Baux) highlights this first sort of blink. It tends to leads to what might be called *instrumental intuition*. Instrumental intuition is likely to predominate in strongly sensate scientists whose interests lie in harnessing as well as exploring the physical world. When facing uncharted terrain, they may be especially likely to use metaphors grounded in embodied experience to structure their understanding of the external world.[21] Most likely they do not realize they are using a form of intuition; rather they think they are objectively describing the world.

In the other mode, the *outside-in* mode (the one more often considered intuition), the opposite tends to occur. Intuitive thinkers are more likely to interject knowledge of the external world gleaned from unconscious observation and integration inward into physical or visceral sensations or into embodied imagery. They are then particularly attuned to the subtle subjective kinesthetic patterns they express.

Embodied experience is a particularly effective medium for transferring unconscious knowledge to the conscious mind. It is likely to be much more spontaneous and true than other kinds of knowledge because it is not directly available to the conscious mind, with its established ways of thinking and its tendency to revise more direct experience. Here the intuitive *blink* is again an instant of blindness as well as creativity, but one that transforms *the eye* into *the I*. It results in what I think of as *receptive intuition*. With receptive intuition, information about the external world is received in *I-space*. The blind transformation between the *inside* and *outside* inherent in these two modes of perceiving the world poses a considerable challenge to any facile notions of objectivity and subjectivity, as does so much of the material in this and other chapters.

## EMBODIED IMAGERY IN THE WORK OF UNUSUALLY INTUITIVE SCIENTISTS

Newton demonstrated that the same laws govern the motion of earthly and of heavenly bodies. The Newtonian universe can also be seen as an externalization, or projection, of bodily experience onto celestial as well as other earthly bodies. Of course Newton's laws rested on past empirical observation. However he clearly experienced the laws of mechanics at an implicit level through his body long before he studied Kepler's work.

Until the early 20th century, the assumptions behind Newtonian physics remained unquestioned. Absolute space, absolute time, and mechanical causality are inherent in our experience of embodiment; they come to form our commonsense ideas of reality. Physicists had no reason to consider that these notions might have more to do with the level of physical reality our bodies evolved to cope with than with universal law.

These notions are also the preconditions for rational thought. It is not easy to move beyond them, and for most purposes there is no need to. The 20th century physicists whose work pushed beyond their limitations necessarily depended on them to a large extent to structure their thinking. To illustrate the power of body-based experience and imagery to capture profound unconscious understanding, I consider two 20th-century scientists who had highly intuitive cognitive styles. Both Albert Einstein and David Bohm depended unusually strongly on embodied experience in their work, and both were able to use their intuition to transcend the limits of the embodied worldview of Newton.

### Albert Einstein

Einstein's work in the early 1900s superseded the absolutes inherent in Newton's universe. Einstein's thinking was thoroughly suffused with embodied imagery. He used it to push commonsense notions to their limits to reveal their potential contradictions. In what might be the most startling embodied image of all times, as a 16-year-old he pondered what it would be like to ride on a beam of light.[22] Einstein credited his fanciful adolescent image with encoding the germ of the theory of special relativity. It asks, "If the speed of light is the fastest speed allowable, how would light itself seem to travel for someone themselves moving at the speed of light?"

This image as well as the thought experiments he later used to crystallize and communicate the special and the general theories of relativity show the limitations of body-based notions by stretching them to take into account a larger cosmos in which these views create cognitive paradoxes. To explain the special theory of relativity, for example, he used a thought experiment of two flashes of light that are simultaneous for a stationary observer but not for one that is moving. If, as he proposed, the speed of light is invariant in all frames of reference, then absolute notions of time and space and simultaneity do not work for observers moving at high speeds with respect to each other. He asks us to expand our body-based perspective by inviting us to become several different observers whose frames of reference do not jibe with each other. These differences lead clocks to slow down relative to each other and measuring devices to shrink.

Our first learning is bodily learning. It is fitting that the initial steps in restructuring commonsense notions derived from this experience would also work with the body's own habitual understanding. Arguably only someone like Einstein who was able to touch with relative ease the embodied experience behind these commonsense perspectives could break through to this wider view of the universe.

Einstein's thought processes had their beginnings in inarticulate visual and muscular signs. In response to a questionnaire by Hadamard about the kind of mental images (or the kind of "internal words") his mind used, he wrote:

> *The words of the language, as they are written or spoken, do not seem to play any role in my mechanism of thought. The psychical entities which seem to serve as elements in thought are certain signs and more or less clear images which can be "voluntarily" reproduced and combined. … There is, of course, a certain connection between those elements and relevant logical concepts. … The above mentioned elements are, in my case, of visual and some of muscular type. Conventional words or other signs have to be sought for laboriously only in the secondary stage, when the mentioned associative play is sufficiently established and can be reproduced at will.*[23]

Thought for Einstein manifested in I-space as inarticulate internal sensory or motor signals that had intuitive meaning for him and he could "voluntarily" manipulate. Yet during at least certain moments of intuition, such as when he had the image of riding on a light beam that he credited with containing the germ of the special theory of relativity, his imagery was much more concrete and external—albeit in a rather unusual way. This illustrates once again that sometimes the least used way of experiencing at the conscious level carries unconscious knowledge to awareness. Einstein's vibrant and dashing image, like my experience at Les Baux, captures the freshness and exuberance of the least-favored mental function. However, what is even more important in this context is what Einstein's usual thinking mode shares with this extravagant image, as well as with the thought experiments that framed his breakthrough understanding. They all had a higher than usual sensorimotor content, which places the experience of the participant/observer rather than conventional signs or images at their core.

## David Bohm

Einstein did not accept the randomness enshrined at the core of the accepted interpretation of quantum mechanics, famously saying, "God does not play dice with the universe." He suggested instead that *hidden variables* governed quantum events. David Bohm, a younger colleague, likewise became distressed that quantum mechanics, although an effective method for calculating verifiable results, had no coherent theoretical basis. He developed a mathematically valid theory in which *nonlocal* hidden variables generate a quantum potential, which in turn determines quantum events.[24] The quantum potential guides particles holistically rather than mechanically.

Over the course of his career, Bohm developed a new understanding of physics in which the apparent world of separate objects, space, time, force, and mechanical causality, what he called *the explicate order*, is enfolded within a deeper level of reality, *the implicate order*.[24] Like a holograph, the tiniest part of the implicate order contains information about the whole. Such a radically revamped understanding of the cosmos can account for some of the more unusual experimental findings in quantum mechanics, such as the nonlocality suggested by quantum entanglement in which one particle can influence another widely separated particle instantaneously. (Bohm actually had a role in the series of events that led to the demonstration of nonlocality.)

Bohm's ideas were not well accepted by the physics community at the time. However Einstein and Feynman were enthusiastic about his views. Now called Bohmian mechanics, his work is enjoying a slowly but steadily increasing popularity.[25]

In addition to quantum entanglement, his thinking can account for the nonlocal view of consciousness embraced by some cognitive scientists, based in part on their own work.[26] Einstein's revolutionary insights transformed the physics of the 20th century; Bohm's radical understanding of underlying reality (or perhaps some other holistic theory) might transform the physics of the 21st century.

Bohm used his body to understand the dynamics of physical systems, and only later used mathematics to formalize his thinking.[24] Gyroscopes present a challenge to our sense of the way things ordinarily move. When pushed, they go at right angles to the direction common sense tells us they should, and they sometimes also seem to defy gravity. Most students studying gyroscopic motion are satisfied with explanations about the conservation of angular momentum in conjunction with learning the appropriate formula. However, this did not fulfill Bohm's urge to understand the phenomenon. He needed a more intimate physical sense. Like the inventor in Chapter 6 who took the problem of inventing a new kind of altimeter into his body by pretending he was attached to a huge spring, Bohm, on a walk one day, imagined himself as a gyroscope. By some sort of interiorization he was able to get an embodied sense of its motion, which is what he needed to comprehend gyroscopes.

Likewise, he used his body to understand the counterintuitive notion in quantum mechanics that the combination of two equal and opposite spin states produces spin in a new dimension.[24] Through a sense of equal and opposite tensions in his body, he experienced himself spinning up and spinning down. Then suddenly the two came together and transformed into something else that was qualitatively different.

Bohm explicitly assigned his body the role of capturing abstract unconscious insight. He saw himself using it as a lens-like instrument that could gather scientific understanding of the cosmos, consistent with his holistic view of the cosmos in which each part enfolds the whole.[23] His capacity for embodied understanding, whatever its source, demonstrates its potential to convey complex abstract insights that transcend bodily experience itself.

In their own way, both Bohm and Einstein used embodied experience to concretize their abstract insights about the universe and bring them to awareness. They were also able to translate unusually clearly between this sensorimotor knowledge and conscious understanding. Along with their exceptional intelligence, this allowed them to recognize the blinders that our embodied perspective imposes on our view of reality and to suggest theoretically sound enlarged versions.

## REFERENCES


1. Lakoff G. *Women and fire and dangerous things: what categories reveal about the mind.* Chicago: University of Chicago Press; 1987.
2. Johnson M. *The body in the mind: the bodily basis of meaning, imagination, and reason.* Chicago: University of Chicago Press; 1987.
3. Mervis CB, Rosch E. Categorization of natural objects. *Annu Rev Psychol* 1981;**32**: 89–113.
4. Jung CG. *Psychological types.* Princeton N. J: Princeton University Press; 1971. [Hull RFC, Trans.].
5. Wills WT, Soraci SA, Chechile RA, Taylor RA. "Aha" effect in the generation of pictures. *Mem Cogn* 2000;**28**:939–48.
6. Tettamanti M, Buccino G, Saccuman MC, et al. Listening to action-related sentences activates fronto-parietal motor circuits. *J Cogn Neurosci* 2005;**17**:273–81.
7. Hauk O, Johnsrude I, Pulvermuller F. Somatotopic representation of action words in human motor and premotor cortex. *Neuron* 2004;**41**:301–7.
8. Nazir TA, Jeannerod M, Hauk O. The role of sensory-motor systems for language understanding. *J Physiol Paris* 2008;**102**:1–3. [This is an introductory article to a special issue of the journal devoted to embodied semantics].
9. Glenberg AM, Kaschak MP. Grounding language in action. *Psychon Bull Rev* 2002;**9**: 558–65.
10. Gallese V, Lakoff G. The brains concepts: the role of the sensory-motor system in conceptual knowledge. *Cogn Neuropsychol* 2005;**22**:408–18.
11. Desaii RH, Binderi JR, Conat LL, Mano QR, Seidenberg MS. The neural career of sensory motor metaphors. *J Cogn Neurosci* 2011;**9**:2376–86.
12. Gallese V, Goldman A. Mirror neurons and the simulation theory of mindreading. *Trends Cogn Sci* 1998;**2**:493–501.
13. Rizzolatti G, Sinigaglia C. The mirror mechanism: a basic principle of brain function. *Nat Rev Neurosci* 2016;**17**:757–65.
14. Kilner JM, Lemon RN. What we know currently about mirror neurons. *Cell* 2013;**23**: R1057–62.
15. Gallese V. A neuroscientific grasp of concepts: from control to representation. *Philos Trans R Soc Lond B Biol Sci* 2003;**358**:1231–40.
16. Hickok G. *The myth of mirror neurons: the real neuroscience of communication and cognition.* New York: Norton; 2014.
17. Donald M. *Origins of the modern mind: three stages in the evolution of culture and cognition.* Cambridge (MA): Harvard University Press; 1991.
18. Zhong CB, Liljenquist K. Washing away your sins: threatened morality and physical cleansing. *Science* 2006;**303**:1451–2.
19. Zhong CB, Leonardelli GL. Cold and lonely: does social exclusion literally feel cold? *Psychol Sci* 2008;**19**:838–42.
20. For a study showing that implicit learning can be transferred metaphorically in a bidirectional manner see Guo X, Li F, Yang Z, Dienes Z. Bidirectional transfer between metaphorical related domains in implicit learning of form-meaning connections. *PLoS One* July 2013;**8**(7):e68100. https://doi.org/10.1371/journal.pone.0068100.

21. Fox-Keller F. *Refiguring life: metaphors of twentieth century biology*. New York: Columbia University Press; 1995.
22. Holton G. *Einstein, history and other passions*. Reading MA: Addison-Wesley; 1996. p. 199.
23. Einstein A, quoted in Hadamard J. *The Mathematician's mind: the psychology of invention in the mathematical field*. Princeton (NJ): Princeton University Press; 1973. p. 142–3.
24. Peat DF. *Infinite potential: the life and times of David Bohm*. New York: Addison-Wesley.
25. Talbot C. Recognition of Bohm's causal interpretation. In: Talbot C, editor. *David Bohm: causality and chance, letters to three women*. Oxford UK: Springer; 2017.
26. Forman RCK. A watershed event: neuroscience, consciousness, and spirituality. *J Conscious Stud* 2008;**15**:110–5.

# CHAPTER 8

# A Feeling for the Truth

## Contents



A feeling for the truth is often viewed as synonymous with intuition. However, it is only one component of the evaluative side of intuition. Feelings of truth along with their close relatives such as feelings of rightness and feelings of knowing also fall under the rubric of *metacognition*, the name given to the monitoring and feedback mechanisms that guide and control cognition as it occurs.

Consider the sentence: The sneakers had to be thrown out because the sign blew away. Then think *newly poured concrete*.[1] The feeling of something being wrong turns into a feeling of rightness. It is also a feeling of knowing in the sense of knowing what the sentence is about.

William James, one of the first to describe these generally elusive feelings, captured the core of what is probably the most dramatic kind of feeling of knowing, the tip-of-the-tongue experience:

> *Suppose we try to recall a forgotten name. The state of our consciousness is peculiar. There is a gap therein, but no mere gap. It is a gap that is intensely active. A sort of wraith of the name is in it, beckoning us in a given direction, making us at moments tingle with the sense of our closeness and then letting us sink back without the longed-for term. If wrong names are proposed to us, this singularly definite gap acts immediately as to negate them. They do not fit into its mold. And the gap of one word does not feel like the gap of another, all empty of content as both might seem necessarily to be when described as gaps.*[2]


*Understanding Intuition*
ISBN 978-0-12-814108-3
https://doi.org/10.1016/B978-0-12-814108-3.00008-7

As the tip-of-the-tongue experience suggests, feelings of rightness or knowing can report on information even if it is not available to awareness. Likewise, in an early study of intuition, Kenneth Bowers[3] found that subjects could tell at better than chance levels whether three words, such as *base*, *snow*, and *dance*, have a common associate even though they could not come up with the linking word. In contrast to the compelling tip-of-the-tongue experience, however, Bower's subjects did not necessarily have a conscious metacognitive experience. Their answers were informed by metacognitive information that may have been conscious, unconscious, or somewhere in between. (The common associate is *ball*.)

What we take to be feelings of knowing and rightness are sometimes wrong. They can also be manipulated by a variety of situational and experimental conditions.[4] Partly because of this, many researchers have argued that metacognitive feelings do not directly monitor hidden memory traces or hidden information processing. Instead they view them as inferences based on indirect information, such as familiarity, the fluency or ease of processing, and/or the number of associations that are evoked.[4–6]

The view that feelings of rightness can sometimes reflect direct monitoring of hidden information has also had its academic champions.[7] Metcalfe,[7] for example, pointed out that aphasic patients can *know* the word they cannot say because they can describe it in a roundabout way, or use pantomime to get others to say it and then confirm that it is what they mean. She also pointed to people with Capgras disease, who feel that family members are impostors, even though they look and play the part, to argue that something in addition to inferential knowledge is necessary for a real feeling of knowing.

What is familiar, what comes to awareness quickly, or evokes the most associations may feel right whether or not it is so. The notion that evaluative feelings necessarily depend only on indirect information is a good fit with the view that intuition is primarily habitual and/or heuristic thinking that occurs below the level of awareness and may or may not be right. However it meshes less well with the view of intuition championed here—as a cognitive capacity that allows the mind to move beyond its habitual ways of thinking to important novel understanding. Higher-level intuition sometimes also happens automatically. However, in contrast to habitual knowledge or indirect inference, it orients toward the truth by integrating more and more complexity. The vignette I present in the next section lends support to the view that evaluative feelings that guide this integration can directly monitor hidden information.

When metacognitive feelings that depend on direct monitoring of unconscious contents break into awareness, they tend to have a more refined, impersonal, or objective quality than inferential evaluative feelings. They are sometimes called *noetic* feelings—from the Greek word for knowing, *noesis*.[8] The inventor mentioned in the final section of Chapter 6, for example, noted the especially subtle and impersonal nature of his experience just as he came to the solution. It did not feel like his creation. Rather it lifted from the personal realm and became something about the world.[9]

Much of the time metacognitive feelings register at a conscious level only if we pay attention to inner experience. Therefore it is possible to mistake the more familiar and facile inferential evaluative urges for the more rare and trustworthy noetic ones. This is one reason why what we take for feelings of rightness often seem to lead us astray.

## METACOGNITIVE EMOTION

Emotions mark what has value or matters to us. Metacognitive feelings are emotionally-colored cognitive evaluative signals that convey a gentle, physically-tinged sense of well-being or discomfort—of goodness or badness—in association with ongoing cognition.[10] Whether they are conscious, unconscious, or on the fringe, their emotional tone helps bias cognition in accord with their message.

In Chapter 5 on emotion, I recounted how I experienced a passionate voice that advised me one morning where to pick up my writing as cognitively compelling. Yet the evening before, I had given no credence to a flat voice devoid of emotion that had advised me to do the same thing. The subtle emotive quality associated with noetic feelings tends to fall in between. Although more compelling than the disembodied tone parodying the "voice of reason," it is less compelling, biased, and personal than the passion carried by the emotional tone. As the vignette in the next section illustrates, these subtle feelings of rightness can indeed sometimes have a reason-like role. They can let us know whether our ideas or plans are consistent in themselves, consistent with other information available to us, and consistent with our ultimate goals.

Unfortunately, the louder and cruder force of ordinary emotion readily trumps the subtler and more refined metacognitive feelings. This is another reason why feelings of rightness often seem unreliable. We misread emotional feelings for the more trustworthy metacognitive ones.

The gentleness and subtleness of most noetic metacognitive feelings can explain the paradoxical finding that sometimes the intensity of an *apparent* feeling of knowing or rightness is inversely related to its validity.[11] Emotional excitement about an idea or project can be interpreted as a sign of its impending success. Yet if it is not supported by more impartial feelings of knowing, it may sometimes signal the opposite. The tension between the need for apparent success and closure on the one hand and the quieter metacognitive knowledge on the other can sometimes express itself as emotional excitement—for instance, that the apparent late-hour breakthrough excludes important information and will not hold up in the morning.

By the same token, much of the ability of metacognitive emotions to convey the validity of intuitions comes from their more impartial feeling than ordinary emotions. At some level, we recognize that the information they report is trustworthy, rather than a current best guess, because of their refined and impersonal quality. As the inventor's feelings as he was coming to the solution suggest, these subtle evaluative feelings sometimes seem as if they have come from outside the self, reflecting in part an evaluation made from a deeper place than the conscious mind. They may feel as objective as any information that comes through the senses from the outside.

I have emphasized the potential conflict between ordinary emotion and noetic metacognitive emotion in part to isolate the experience of the latter. Although they often oppose each other, they also can work together. From an evolutionary perspective, the potential for cooperation between ordinary emotion and its more impersonal metacognitive cousin is essential. The enhanced motivation that occurs in emotionally charged situations involving real danger can make signals that generally occur below awareness more likely to emerge closer to consciousness. Also because ordinary emotion can heighten alertness and focus, it can make us attentive to fainter metacognitive evaluative signals. As a consequence, one place to catch a glimpse of these subtle feelings may be in the midst of an emergency situation in which there is little room for error.

## A GLIMPSE OF THE UNCONSCIOUS AT WORK

My understanding of the evaluative aspect of intuition began with an unusual personal experience. This surprising encounter with my unconscious, like "the blink of the eye/I" experience, has had a pivotal role in my understanding of intuition. It highlights that feelings of rightness that perhaps most often occur below awareness can have an extended role in insight formation and intuition.

I had been struggling for a number of years to understand Carl Jung's challenging concept of the *collective unconscious*. At different times I had understood different things he had written about it; nonetheless a unified sense of his concept eluded me.[12] One evening in a class on Jung's work, the instructor made a point about the collective unconscious that struck me as more penetrating than my three other bits of understanding, which I had seen as unrelated to each other. I remember wondering whether the new insight was consistent with these older pieces of information. I tried to recall them but I could not.

Then without willing it, just below my awareness I found myself checking this new understanding to confirm its consistency with this earlier information I could not retrieve but had apparently stored in memory. My link with this process was a feeling trace at the threshold of awareness that recorded the motion between the various pieces of stored information and also corresponded in some way to the contents being compared. It reported a continuous sense of harmony, letting me know that everything was checking out. I sensed that I could probably bring the entire experience to consciousness had I wanted to. Yet the experience available to me was too compelling to stop.

At the end of the process, much to my surprise, I nonetheless had a much deeper and resonant sense of Jung's concept available to me at a conscious level. After a few minutes, however, my attention turned to something else. When I checked again, my newly integrated understanding had also receded out of range of conscious recall.

The richness and complexity of this dream-like encounter mirrors the richness and complexity of the unconscious mind. A veritable cognitive fun house, it distorts many of our most basic notions about consciousness. Its suggestion that the feelings we generally do not experience can guide cognition is surprising in itself. Yet what may be even more surprising is its revelation that these feelings can report on memories the conscious mind cannot readily access. Although I could not bring my old insights to consciousness, something outside my awareness could still compare them with my current, more penetrating insight and report at the level of feeling.

The backbone of the experience consists of two cognitive events: the first a sequential, analytic process and then a second, much faster parallel, synthetic one. The analytic part of the experience compared my (then) new understanding for consistency with each of the three old bits of understanding I could not bring to awareness. The synthetic event that followed, which was every bit as unexpected, depended on the same pieces of information. Without my conscious will or awareness, my unconscious mind synthesized the current insight with the three older pieces of information and brought an even more inclusive insight about the collective unconscious to awareness. At the time I understood the generation of this integrated understanding about the collective unconscious as a separate and only loosely related event. In any case, it also soon receded out of awareness.

From our usual perspective, stored information is viewed as almost useless unless it can be brought to awareness. My scientific colleagues with whom I shared the story all expressed dismay at hearing that my newly enlarged understanding of the collective unconscious had also receded beyond recall after a short while. I also felt the same way when I first realized it was gone. Yet my concern turned to admiration for the "slyness" of the unconscious as I began to assimilate one of the many lessons from the experience. Even information no longer available to awareness can function in intellectual synthesis.

The vignette also demonstrates that the unconscious can initiate and guide higher cognitive processes. I did wonder at a conscious level whether the new insight I just had was consistent with my older insights. But I,—in the sense of my conscious mind—could not initiate the process that ensued because I could not remember the older ideas. The encounter provides an illustration of something that echoes through the book. The conscious mind might not always be at the center of higher-level intellectual activity. We will return to this perhaps disquieting notion later in the chapter.

## CHECKING WHAT AND WHY?

Assembling the bits and pieces that make up our insights generally happens out of awareness, so we tend to assume that it happens automatically. Insight is perhaps the best-accepted aspect of higher-level unconscious cognition. Ironically the vignette suggests that the process of combining ideas below awareness may not always be as automatic as it seems. At least when joining a new piece of understanding with older ones, the final synthesis may be preceded by a step outside awareness that compares the old and new knowledge.

One of the most tantalizing aspects of the experience is that the feelings of harmony associated with the first event were a response to a process that even now remains veiled. Just after it occurred, I just assumed that the feelings I experienced during the checking process were a positive answer to the relatively limited analytic question my conscious mind had posed: "Is this new insight about the topic consistent with my older bits of understanding?"

At one level this is a propositional question. It seems to depend on what is generally referred to as the correspondence view of truth in which statements do or do not correspond to what is true in the world, or in this case, in my mental world. Propositional questions can be answered yes or no, and propositional truth is generally associated with the conscious mind.

But was there more going on than this analytic, albeit largely unconscious, exercise? A bit later it struck me that the question my conscious mind had asked about whether the new insight was consistent with the older pieces of information reflected more than idle curiosity about the consistency of these different ideas. Implicit in the question was a desire for a more comprehensive understanding of Jung's elusive concept of the collective unconscious. (I suspect this desire remained below awareness because I had come to assume that an integrated understanding of an idea as elusive as the collective unconscious would always elude me.) In this broader context, I saw that the feelings of harmony also probably reported that such a link would be helpful to understanding the concept: that is, each set of two ideas cohered in a manner that might lead to a fuller understanding. Moreover, I recognized that these feelings might have reported the presence of multiple bridges between each of the older ideas and the newer, more penetrating one.

I now think the feelings of harmony I experienced indicated that each of the three old ideas could be brought into a rich relationship with the current, more penetrating one. This means that the newer insight overlapped with each of the older pieces of knowledge in one or more ways that expanded or refined it without conflicting at a logical level. By linking to each of the three old pieces of information in a meaningful way, the new insight with its greater depth also brought them into a potentially meaningful relationship with each other.

So lurking just inside the overt propositional question was another potential question, one that looks at truth in a different way. In the coherence view of truth, ideas are judged by the way they fit together with other ideas instead of their correspondence with the world. Coherence is in play when we know something is true because its parallel interactive strands come together below awareness to form a meaningful whole. This is the kind of truth that is generally associated with unconscious cognition and thus with intuition. Yet the vignette seems to be cautioning, "Not so quick. Both kinds of truth may have a role in intuition."

Let's take a deeper look at the checking process. To get at the purpose a hidden checking process might serve in intuition and insight formation, consider what might have happened if one of the old pieces of knowledge had not checked out with the new one. I suspect that the discomfort associated with the feelings of wrongness would have brought the old piece of information to awareness. This is based in part on my sense at the time that, had I wanted to, I could bring the whole process to awareness. Support for this also comes from an experience we probably all have on occasion. After hearing or reading something noteworthy, the thought, "But that does not fit with X!" or, "How does that relate to Y?" might just pop into mind. Indeed a study showed that phrases presented to subjects subliminally that do not make sense, such as "I ironed the coffee," are more likely to come to awareness than inconsequential information that does make sense.[13] In the context of insight formation, only negative, not positive, feedback is important for the conscious mind. This in turn suggests why the checking step can occur below awareness.

Consciousness is a limited resource. One of its tasks is to sort out problems that arise from negative evaluative feelings during automatic processing. Activating the relevant memories below awareness and checking them there for consistency with new information allows consciousness to maximize its influence.[14] It would be required only to deal with problems that arise from dissonant evaluative feelings. Had one of the old pieces of information not fit with the newer one, my conscious mind might have decided that the old one was actually irrelevant with respect to the issue or that the conflict between the two was only apparent.

Even more important from the perspective of the movement toward truth, my conscious mind might have recognized an important disjunction between my new insight and one of the older bits of understanding. If synthesis proceeded without checking and without the hidden problem coming to awareness, the resulting insight might have been incomplete or wrong. On the other hand, if checking occurred and a negative metacognitive evaluation prevented synthesis *without* simultaneously presenting the problem to awareness, an opportunity for my conscious mind to bring its resources to bear on what might be an important contradiction or gap in my store of knowledge might have been lost. The vignette points out that a checking step below awareness before synthesis is a good cognitive strategy and one that only takes up the limited resources of the conscious mind if something does not fit.[13] It is a strategy that may have a particularly important role in breakthrough intuitions. The metacognitive discomfort that might result focuses both the conscious and unconscious mind on the fault lines of our knowledge, which is especially fertile terrain for the incubation of intuitions. By illuminating the gaps and potential contradictions between new and old knowledge, the checking process can force the mind to go deeper. It reorients cognitive activity toward what is not yet known or understood. By pointing out what is not known, it reveals what might be known.

Yet clearly checking does not always occur. On occasion, we will accept without question our own or someone else's ideas that are grossly inconsistent with other knowledge we possess. This is especially likely when emotion is strongly involved. Moreover personality differences may influence whether checking occurs. Studies suggests that some people may be more motivated by avoiding errors than others.[15]

In certain situations, "not checking" is an advantage. As social animals, our intelligence is geared as much to convincing others about what we want them to do or think as it is to the pursuit of an increasingly deeper understanding of the world.[16] Deception and oversimplification can have an important part in convincing others of one's own point of view. The ability to deceive others is made easier by self-deception. Being able to hold only half the story at a conscious level when dealing with others without being bothered with what does not fit (something that seems to have become a high art form in the hand of certain politicians) may have driven the evolution of the brain almost as much as our feeling for the truth. However, as the next chapter emphasizes, this may no longer support us as a species.

## A PRIVILEGED VIEW OF THE INTERACTION OF THE CONSCIOUS AND UNCONSCIOUS MIND

The vignette illustrates the layered nature of the unconscious and also that what is usually hidden below awareness may on occasion be revealed. The feelings of rightness that appeared in the center of my awareness summarized a checking process I could partially follow in the fringe of awareness, which called on specific memories I could not access. The feelings accompanying the veiled checking process seem to occur in a part of the unconscious that is close to the interface with the conscious mind. In an experience in which I had much more access to hidden processes than usual, these feelings were highlighted. The checking process itself seemed to happen at a lower level in direct contact with the activated memories. Yet through my feelings I could still experience some of the movement that occurred during the process. The final synthesis of all four pieces of information seemed to have happened at an even more remote level; I did not have even indirect access to the integration. This layered view of the interaction between the conscious and unconscious mind is consistent with the idea that the divide between them is a continuum and, moreover, that where any cognitive event appears on the continuum is responsive to many different factors.

The layered view of the unconscious mind suggested by the vignette also dovetails nicely with Bruce Mangan's[17] idea that feelings of rightness act like the pull-down menus that appear at the top of computer programs. They report on information relevant to the topic that is not currently available to consciousness. This allows consciousness, like the computer screen, to make the most of its limited display capacity by showing only the most important information. Just as menus can call forth other menus on a computer screen, there is a complex hierarchy of linked layers in the unconscious. Through such a cascade the conscious mind can receive information about increasingly less accessible processing and memories. In this manner, feelings of rightness can reflect the coherence of information currently displayed in awareness with a whole range of information that is unavailable to awareness.

## A MENTAL BUFFER MEDIATING BETWEEN THE CONSCIOUS AND UNCONSCIOUS MIND

This unusual experience suggests the presence of a multilayered mental buffer generally hidden from awareness that shares certain characteristics with the conscious mind and certain with the unconscious mind and seems to mediate between the two. Like working memory, it can hold a number of pieces of old information and compare them with a new piece in a serial manner. Yet unlike the usual view of working memory, which limits it to conscious information,[18] this buffer can work with information that is unavailable to consciousness.[19] The vignette also suggests that the buffer can use hidden information in pursuit of goals that the conscious mind did not set—i.e., synthesizing the three older pieces of information together with the newer one to make a more inclusive insight. It can also keep the final synthesis at bay while it performs its careful check.

This unusual experience also illustrates that movement between the conscious and unconscious mind is bidirectional, a potential two-way street. What is usually hidden below awareness may on occasion be revealed. What is usually retrievable (or emerges closer to awareness) can be enfolded back into the unconscious and become inaccessible—and not just because of the forgetting that comes with time. What begins as a conscious query, such as "Is this insight consistent with my old bits of knowledge about the subject?" can be taken over by an intent that is hidden from the conscious mind. Activated memories can be checked at a level that is at least partially available to awareness and then put together in a process that is not only opaque to the conscious mind but also unbidden and unexpected by it. Moreover penetrating insights such as the final insight of the evening that was transiently available to awareness can disappear into the unconscious. The vignette illuminates the complex and sinuous dance that can occur between the readily accessible and generally inaccessible aspects of mind. It also highlights the important role that forgot information can have in insight formation and intuition.

## THE INTUITIVE SELF

A key question is "who" or "what" choreographed this retrograde movement, and indeed the entire unusual experience. Who or what "decided" that the activated memories should remain below awareness, that a metacognitive check should occur and be partially revealed to awareness, that further synthesis should occur, and perhaps even that the final integrated insight should disappear from awareness?

The surprise we tend to experience about this "retrograde" movement of agency and insight reflects our sense that consciousness conducts our mental symphony. It is higher in the sense of being more integrated, more systematic, and epistemically richer than the unconscious mind. Movement is always upward, except for the forgetting that occurs with time.

Yet what seems like a takeover by the unconscious mind in the vignette may instead reveal what is generally true. The default organizing center of our experience may not be the conscious mind, but an intuitive, or implicit self that resides below awareness—a proposal consistent with much of the other material in this book. The intuitive self controls behavior and directs experience from below awareness in an automatic, although purposeful, or intentional, way.

The intuitive self determines our perceptions below the level of awareness and most of the time hands the conscious mind a ready-made meaningful world. It sets our behavior in motion based on a series of nested goals ranging from short to long term, informing the conscious mind largely on a need-to-know basis. In alliance with a cognitive buffer also generally hidden from awareness, it can evaluate old information in light of new and, when necessary, issue interim metacognitive reports. Much of the time all goes well, and the larger, more inclusive insight that might result, like most of our other thoughts, just pops into mind.

The intuitive self integrates the moment-to-moment opportunities offered by the environment with our past experience and genetic proclivities to determine our goals, as Chapter 4 on the brain suggests. Most of the time, the goals of the intuitive self jibe with those we can report at a conscious level. They support our usual sense of agency and "I-ness." We tend to see evidence for the determining role of the hidden self only when conscious and unconscious goals are in conflict, as may occur with neurosis, with posthypnotic suggestion, in "senior moments," and in certain experiences of intuition, such as the one described by the vignette in this chapter.

Decades of research suggest that people are often unaware of their motives, whether they are triggered by internal or external factors.[20] A large body of work in the social sciences reveals that external or situational factors can motivate goals below awareness.[21] A study showed that even the *subliminal* promise of a higher monetary reward resulted in more forceful hand grips (a measure of behavior activation) than a lower reward.[22]

The demonstration that external forces of which we are unaware can motivate behavior bolsters the case that internal needs we are unaware of can as well. In most animals goals are not represented at a conscious level, nonetheless their survival depends on intentional behavior. Even for humans, consciousness is a relatively limited resource, which means the unconscious mind/intuitive self must set in motion most routine behavior. The work of Libet and others[23,24] points to the possibility that voluntary behavior may also have its roots in unconscious signals that precede the conscious urge to act. It is unlikely that the conscious mind could keep track of the many layers of nested goals that help choreograph the present moment and at the same time give full attention to the most demanding one.

In addition to initiating behavior, an intuitive self would have to perform complex metacognitive monitoring and control operation below awareness. Research on judgment and decision making increasingly recognizes that metacognitive monitoring and control processes, which were previously thought to be the exclusive property of consciousness, can occur unconsciously.[25] For example, studies have shown that metacognitive signals in response to errors that subjects are unaware of making lead them to correct their responses.[26]

Subliminal stimuli can also activate the prefrontal cortex and initiate inhibition. In another task, called a go/no-go task, the subject has to respond to a cue when it is preceded by the go-signal, a circle. However, when the same cue is preceded a triangle, the no-go signal, the subject is supposed to refrain from responding.[27] Imaging studies show that the prefrontal region reacts in the same way whether the no-go signal preceding the cue is subliminal or conscious. The no-go signal activates the inferior frontal cortex, which is part of what has been called *the inhibition network*. However with strongly masked no-go signals, the response tends to be slowed rather than inhibited.

Finally, a follow-up study to the experiment mentioned in the beginning of the chapter is noteworthy because it evokes the potential power of unconscious information processing at least in the associative realm. It showed that solutions to difficult remote association triads can prime subsequent cognitive activity. This means that common associates for coherent triads that cannot be solved are nonetheless activated below awareness.[28] (Intriguingly the data also hint that the priming ability of solutions to difficult problems that do not eventually appear in awareness might even be stronger than for those that do.)

There is, of course, a long way to go to show convincingly that an intuitive self conducts our mental symphony. However, even these few examples from the available evidence, when taken together, suggest that the unconscious mind directed by an intuitive self is capable of complex information processing and cognitive monitoring below awareness, and that it can initiate as well as inhibit behavior. Its ability to process information that is unavailable to awareness is probably much more powerful than we can appreciate, given our reflexive identification with our conscious knowledge and conscious intelligence, and our still limited albeit growing experimental strategies. The vignette hints that the different abilities of the nonconscious mind may combine in ways that surprise us—including those of us are already convinced of its cognitive prowess.

The potential agency of the unconscious mind and intuitive self does not reduce the conscious mind or self to a ceremonial role. The troubleshooting ability of consciousness is essential to adaptive functioning, and consciousness also has veto power over voluntary action.29 It excels at certain kinds of cognitive skills such as planning and resolving apparent contradictions, and it may be crucial to framing hypotheses. Also focused conscious attention can greatly enhance the power of unconscious attention and intent. Finally, consciousness allows for amplification of the experience of pleasure. It may be required for what is arguably one of the subtlest yet profound kinds of pleasure humans can experience: the feeling of truth.

## FEELINGS OF TRUTH

How we recognize the rightness or truth of our insights is the deeper meaning of the title of this chapter, "A Feeling for the Truth." The vignette illustrates at least three different kinds of feelings of rightness that can participate in the unconscious construction and evaluation of insights and intuitions.[30]

The first kind reports whether specific information in memory, which may or may not be conscious, fits with more current information or other information in memory. The *feeling of relevance* or *fitting* can probably occur completely below awareness. It may report only yes or no, and most of the time it automatically brings what does not fit to awareness. The second is the feeling of illumination and penetration that accompanies the experience of insight itself. The third, which is implicated more by its absence than presence in this experience, is the feeling of truth associated with intuition at its most profound. The three make progressively more stringent demands. I will focus on the second and third kinds of feeling of rightness. The first, the feeling of relevance, was considered in a previous section.

### Feelings of Illumination and Penetration

Our insights often appear in the conscious mind as givens and have the same sense of immediacy as our sensory perception. We take what we perceive to be valid knowledge about the world, at least for the moment. Like visual perception, the perception is composed of interlocking strands that come together below awareness to form a meaningful whole. For this reason, the experience of insight has been referred to as "seeing that." This inner seeing is, of course, one of the many meanings of the phrase a "blink of the *eye*."

The metacognitive feeling of knowing that accompany moments of "seeing that" tends to be overshadowed by the insight itself, which quickly occupies the center of awareness. Nonetheless we can sometimes sense it as a quicksilver flash of illumination on the fringe of awareness, the proverbial light bulb above the comic strip character's head, just as all relevant information coalesces below awareness. This feeling is probably easiest to capture when something else demands attention (such as a ringing phone or a crying baby) and the insight is lost. All that remains is a memory of the feeling of illumination.

The feeling of illumination can have a range of intensity. When it is particularly intense, it may appear in awareness as a sense of penetration or depth that shares billing in consciousness with the insight itself. The insight that initiated the vignette, for instance, was accompanied by a sense of being deeper than my three older bits of knowledge, and the final insight that synthesized all four ideas, much deeper still. The feeling of penetration or depth that sometimes accompanies an insight provides one measure of its cognitive coherence and how well it fits with all the other information in the system—thus its explaining power. The more penetrating the insight, the more encompassing it is, and the more efficiently it can organize all thought, not only at the conscious level but at an unconscious level as well. Thus the less likely it will be overturned.

Yet no instant of "seeing that" is necessarily permanent. A feeling of metacognitive discomfort might begin to overshadow the sense of comprehension and closure associated with the insight. The goals of the intuitive self are constantly readjusting. What seems like a worthy insight in one moment may need amending in the next. The momentary field of coherence that comes into focus in an instant of knowing might include only a limited amount of our potentially relevant conscious and unconscious information. Also, the limitations of an insight may emerge over time as some of its subtler implications rise to awareness.

## Truth and a Sense of Wholeness

Feeling that an insight is valid, or still valid, can be different from a feeling of truth. The feeling of knowing that characterizes intuition at its most profound is the result of a much more exacting process. The word "truth" is used to characterize an insight that encompasses the circumference of a phenomenon at the same time that it penetrates to its core. Thus, although I experienced the comprehensive insight at the end of the vignette that wove together all four realizations about the collective unconscious as much deeper and more coherent than the earlier one, I did not have the sense that even this enlarged perspective fully captured Jung's concept of the collective unconscious—never mind actually representing something true about the world.

Although penetrating and rich, it did not feel complete or whole. Unlike the inventor's solution for the altimeter mentioned in the beginning of the chapter, it did not transcend the subjective intellectual realm and take on a life of its own. However my understanding of my experimental data from isolated lysosomes described in Chapter 3, in contrast, did come with a feeling of objectivity, of being something about the world. This was reflected in my sense of respect for these experimental results and their separateness from me, which in turn created an enhanced experience of grounding and of self—of Presence.

Barbara McClintock[31] in a statement to her biographer, quoted in Chapter 3, spoke about intuition's logic of the whole. She knew an idea was true when it rested on a number of pieces of evidence that were sufficiently interlocked so as to exclude all ambiguity or other ways of understanding. Because of their pattern of interconnection, they point beyond themselves to a *unique* whole.

I trust that McClintock was thoroughly rigorous in her conclusions. Yet I believe that something *more* than an apparent whole at a conscious level is necessary to generate a feeling of truth. This something more is implicit in McClintock's understanding of why she could be so certain of her thinking in certain cases. To generate a feeling of truth, the conclusion must fit with everything we know at an *unconscious* as well as at a conscious level. Unless our understanding is backed by subtle metacognitive signals from the unconscious, it may well prove wrong. The title of Evelyn Fox Keller's biography of McClintock, *A Feeling for the Organism*, evokes the subtle metacognitive emotive signal that announces this wholeness and points to even more complexity and coherence than the conscious mind can necessarily comprehend.

## Can a Feeling of Wholeness Be Wrong?

Just because we might experience a refined metacognitive feeling that says that the pieces of evidence fit together in such a way as to make a unique whole, does this mean that this particular understanding is necessarily true? Because an intuition generally rests on a number of distributed but intertwined strands, it has a certain amount of inherent flexibility. For example, an aspect of Pauling's initial insight about sickle cell anemia proved wrong—that the sickle-shaped cells were distorted by ordered, rigid structures of hemoglobin. However this did not affect the core of the insight.

Yet given a powerful enough correction, an intuition might become unmoored. New information might generate scenarios not imaginable now that would fit better, and this is especially likely if the old information has been somewhat revised. Metacognitive feelings of truth, no matter how objective and refined they may seem, still point only to relative truth.

The existence of understanding by the unconscious, even though it does not directly contradict this view, suggests that metacognitive feelings might still be a good indication of the staying power of an idea. The unconscious knows more than the conscious mind. It can also recognize subtle internal connections between different levels of an insight that give it an inner resilience that would be extremely unlikely, if not impossible to overcome. This tacit information, moreover, may suggest deeper levels of understanding and interconnection that the conscious mind cannot possibly imagine. I propose that feelings of truth indicate that the insight integrates the relevant information so that it works at all possible levels of current conscious and unconscious understanding. A feeling of truth represents a hypothetical end point in the feeling of penetration discussed earlier. When an insight organizes all relevant conscious and unconscious information the person has so that it forms a completely self-consistent whole, we experience it as true.

We sometimes speak of an idea as a bell "ringing true." If we listen with complete receptivity to a fine bell, we can sense its continuing vibrations for a while after it goes out of range of hearing. Likewise, we can sense the coherence of a profound insight as it resonates from deeper and deeper in the unconscious. As it recedes, it carries aspects of awareness with it and exerts an increasingly subtle and refined metacognitive pressure on consciousness.

The high degree of coherence associated with an insight that evokes a feeling of truth makes it unlikely to be overturned by new concepts or new data. By the same token, it may well predict whole classes of phenomena that have not yet come to light. It still might be superseded by a newer more inclusive theory that can penetrate even deeper and encompass even more. However the original insight will nonetheless likely maintain the coherence associated with it and continue to function well in its sphere.

A story about Einstein in the early days before the special theory of relativity was confirmed by empirical evidence illustrates the power of such coherence. The results of an experiment by another physicist convinced many that the theory was wrong. Although Einstein could find no errors in the other physicist's calculations, he simply dismissed the results as flawed.[32] Like McClintock, he was able to appreciate much more coherence than most at a conscious level, and this was no doubt true at an unconscious level as well.

Placing a high value on intuition is sometimes thought to imply a worldview that holds we can perceive absolute truth. Some object to the idea of intuition for this reason. The view of intuition I have presented here is consistent with a worldview that stakes the more modest claim that it is only possible to get closer and closer to the truth. As we continue to explore a phenomenon at both the conscious and unconscious levels, our old understanding may fail to assimilate all we know. This leaves room for a significant expansion in the meaning of the whole, as occurred when Newton's view of the cosmos gave way to Einstein's.

Sometimes unconscious coherence on its own, without conscious understanding, is enough to generate a feeling of truth. When I checked inside a second time to answer my friend's question "What is intuition?" several additional just as incomprehensible images came to mind. Yet because they all encoded the same movement, they convinced me with a certainty that stunned me at the time that "a blink of the eye/I," the silly-sounding phrase that had popped into mind the first time I checked inside, was a valid characterization of intuition. We sense the coherence behind these intuitions sometimes even though our conscious grasp of them is tenuous. Like a bell whose effect on the air can be sensed even when the sound cannot, resonance from the underlying unconscious coherence associated with an idea tends to be transmitted through to consciousness.

## WHY JUNG AND THE COLLECTIVE UNCONSCIOUS?

Recognizing the validity of different approaches to truth is also inherent in the view of intuition I have presented here. In turn, this depends on recognizing that different kinds of underlying coherence are valid. The metaphoric coherence of poetry, the mythic coherence reflected in the rituals of ancient cultures, and even rhythmic and tonal coherence are in some sense as valid as the logical and empiric coherence we associate with scientific truth. Within each of these different approaches to truth, with our unconscious if not our conscious mind, we can register ever deeper levels of penetration as well as recognize what feels true and what does not.

Mythic coherence, which has special relevant to this chapter, finds expression in well-codified metaphors or images that appear in many different cultures. It can give certain ideas that already make sense to the conscious mind a numinous, or "realer than real" feeling. Something about the idea holds a fascination for our mind and continues to draw us. Subtle metacognitive feelings that accompany it as it resonates in the mind suggest that there is even more coherence to the idea than we can currently appreciate.

Kekulé's dream image of the benzene ring as a snake holding its own tail carries this metacognitive feeling of even deeper unconscious coherence. Called the *ouroboros* in mythology, this image of wholeness and unity is an example of one of the many guiding patterns, or archetypes, that Jung posits makes up the collective unconscious. The image also captures the kind of circular logic, or resonating feedback, and also what is referred to as *interiority* in the next chapter, that characterize the complex systems that together make up organic life.

Slowly over time as I have worked on this chapter, I been able to reconstruct the three old pieces of information as well the two new realizations I had about the collective unconscious. The comment that struck me so strongly that evening and initiated the experience was that the collective unconscious encodes the underlying holism of our species. Each of the three previous bits of knowledge about the collective unconscious that I could not recall but remembered as being *unrelated* to each other described one of these realms. The final insight that night recognized these three different realms as different aspects of the same underlying unity.

The three old pieces of information, which I remembered only as being unrelated to each other, were about the archetypes that make up the collective unconscious. These bits of knowledge about the archetypes, in the order that I assimilated them, were: (1) they are source of cultural myths, (2) they are the psychic or mental component of instinct, and (3) they are encoded in the genome and are part of our evolutionary heritage.[33] My final insight that night integrated these three pieces of understanding with the idea of the holism of the collective unconscious to give me a rich sense of the intersection of the biological, the psychological, and the cultural/mythic realm in human experience—of their resonating interconnectivity. The collective unconscious expresses the deep dynamic of the evolutionary experiment of which we are a part.

As penetrating and resonant as the final insight was, it did not take off from the subjective realm and become something about the world. Instead, after only a few minutes, it, too, receded into my unconscious. Why could I not hold on to it? Certainly I had much more to learn about metacognitive feelings, a critical component of the encounter, as well as Jung's concept of the collective unconscious. But perhaps most important, I had no sense of what connected the two.

I have made some progress in the intervening years. Implicit in my insight that night was that the archetypes that make up Jung's collective unconscious provide the unifying themes that fall out of the intersection of the three interrelated realms of human existence. These archetypal themes are the deep structuring grid, the psychic interior, of being human. As the attractors in the human soul, they determine the kinds of meanings we make. The collective unconscious, which occupies a layer of mind less accessible than the personal unconscious, finds expression through intuition as it guides meaning-making and imagination.

The feeling of truth, a powerful force in humans, depends on an unconscious as well as conscious sense of wholeness. It necessarily takes these archetypal themes into account; it seeks a more flexible and adaptive relationship to them. The feeling for the truth drives us toward greater meaning and clarity about who we are, our universe, and what lies beyond.

Jung linked the feeling for the truth to the archetype of Self, which subsumes all other archetypes. Working through a feeling for the truth, the Self strives to unify the conscious and unconscious mind. It seeks to see the world, itself, and the species without the filters that ordinarily frame experience and limit vision. In Buddhists' terms, it is the emptiness that allows the experience of the interconnection of all that is. On rare occasions, conscious and unconscious coherence come together to generate a sense of wholeness strong enough to rise above the realm of ordinary meaning making. For an instant, the subjective transforms into the objective, self feels whole, and we glimpse the truth.

## REFERENCES

1. The sentence was patterned on the test sentences in Auble PA, Franks JJ, Soraci SA. Effort towards comprehension: elaboration or "aha?". *Mem Cogn* 1979;**7**:426–34.
2. James W. *The Principles of Psychology, Vol. 1*. New York: Holt; 1890. p. 251.
3. Bowers KS, Regehr G, Balthazard C, Parker K. Intuition in the context of discovery. *Cogn Psychol* 1990;**22**:72–110.
4. Jacoby LL, Whitehouse K. An illusion of memory: false recognition influenced by unconscious perception. *J Exp Psychol Gen* 1989;**118**:126–35.
5. Reber R, Schwartz N. Effects of perceptual fluency on judgment of truth. *Conscious Cogn* 1999;**8**:338–42.
6. Jacoby LL, Whitehouse K, Koriat A. How do we know what we know? The accessibility model of the feeling of knowing. *Psychol Rev* 1993;**100**:609–39.
7. Metcalfe J. Feelings and judgments of knowing. *Conscious Cogn* 2000;**9**:179–86.
8. Metcalfe J. Feelings and judgments of knowing: is there a special noetic state? *Conscious Cogn* 2000;**9**:178–86.
9. Gordon WJ. *Synectics*. New York: Harper & Brothers; 1961. p. 17.
10. Damasio A. *Descartes' error: emotion, reason and the human brain*. New York: Putnam's Sons; 1994.

11. Metcalfe J. Premonitions of insight predict impending error. *J Exp Psychol Learn Mem Cogn* 1986;**12**:623–34.
12. The course took place at the Boston Jungian Center and focused on the collected works of Carl Jung; the instructor Dr. John Haule is a Jungian Analyst. This vignette first appeared in an abstract for the "Towards a Science of Consciousness Conference 2004" Tucson IV." I had a somewhat similar experience around my scientific work. An earlier publication presented a composite of the two experiences, Isenman LD. Towards an understanding of intuition and its importance in scientific Endeavor. *Perspec Biol Med* 1997;**40**:395–403.
13. Sklara AY, Levy N, Goldstein A, Mandel R, et al. Reading and doing arithmetic nonconsciously. *Proc Natl Acad Sci U. S. a* 2012;**109**:19614–9.
14. There is an additional reason that the kind of checking process described in the vignette might go on below awareness. When new information is encoded by the hippocampus, retrieval of old information from the cortex is thought to be more difficult. Hasselmo M. The role of acetylcholine in learning and memory. *Curr Opin Neurobiol* 2006;**16**:710–15. This is seen as adaptive since it prevents old information from interfering with new. Remembering where the car was parked days before in a large crowded parking lot may hinder the effort to fix in one's mind where it is parked today. Instead consolidation of new and old information tends to occur later during sleep. Yet as the vignette helps highlight, decreased retrieval in many situations may be in tension with the need to determine the quality or relevance of information before it is added it to memory or combined with other information for immediate use. Activating relevant information without having it come to consciousness and checking for consistency below awareness may provide a good compromise, and one that takes into account the limited capacity of consciousness.
15. Mauro Y, Schacht A, Sommer W, Masaki H. Impact of motivations valence on the error-related negativity elicited by full and partial errors. *Biol Psychol* 2016;**114**:108–16.
16. Lauria F, Pressmann D, Clément F. Self-deception as affective coping. An empirical perspective on philosophical issues. *Conscious Cogn* 2016;**41**:119–34.
17. Mangan B. Sensation's ghost: the non-sensory 'fringe' of consciousness. *PSYCHE* 2000;**7**. http//:psyche.cs.monash.edu.au/v7/psyche-7-18-mangan.html.
18. Baars BJ, Franklin S. How conscious experience and working memory interact. *Trends Cogn Sci* 2003;**7**:166–72.
19. Soto D, Silvanto J. Is consciousness need for all working memory? *Neurosci Conscious* July 2016;(1):niw099. https://doi.org/10.1093/niw99.
20. Schultheiss OC, Brunstein J, editors. *Implicit motives*. New York: Oxford University Press; 2010.
21. Bargh JA, Gollwitzer P, Lee-Chai A, Barmdollar K, Troetschel PM. The automated will: nonconscious activation and pursuit of behavioral goals. *J Pers Soc Psychol* 2001;**81**: 1014–27.
22. Pessiglioni M, Schmidt L, Draganski B, Kalisch R, Lau H, et al. How the bran translates money into force: a neuroimaging study of subliminal motivation. *Science* 2007;**316**: 904–6.
23. Libet B. Unconscious cerebral initiative and the role of conscious will in voluntary action. *Behav Brain Sci* 1985;**8**:529–66.
24. Soon CS, Brass M, Heinze H-J, Haynes J-D. Unconscious determinants of free decisions in the human brain. *Nat Neurosci* 2008;**11**:543–5.
25. See, for example, Quartz SR. Reason, emotion and decision-making: risk and reward computation with feeling. *Trends Cogn Sci* 2009;**13**:209–15.
26. Nieuwenhuis S, Ridderinkhof KR, Blom J, Band GP, Kok A. Error-related brain potentials are differentially related to awareness of response errors: evidence from an antisaccade task. *Psychophysiology* 2001;**38**(5):752–60.

27. van Gaal S, Ridderinkhof K, Scholte H, Lamme V. Unconscious activation of the prefrontal no-go network. *J Neurosci* 2010;**30**:414–50.
28. Zhong CB, Dijksterhuis A, Galinsky AD. The merits of unconscious thought in creativity. *Psychol Sci* 2008;**19**:912–8.
29. Libet B. Can consciousness affect brain activity? *J Conscious Stud* 2003;**10**(12):24–8.
30. The vignette also illustrates another kind of metacognitive feeling–a feeling of imminence or nearness to consciousness—the sense I could bring the checking process to awareness, had I wanted to.
31. Fox-Keller E. *A feeling for the organism: the life and work of Barbara McClintock*. New York: W.H. Freeman; 1993.
32. Holton G. *Thematic origins of scientific thought: Kepler to Einstein*. Cambridge (MA): Harvard University Press; 1973. p. 234–5.
33. For one potential connection between genes and archetypes see Spencer HG, Clark AG, Feldman MW. Genetic conflicts and the evolutionary origin of genomic imprinting. *Trends Ecol Evol* 1999;**14**:197–201.

# CHAPTER 9

# Who Are We?

## Contents



The intuitive quest often ends in the spiritual realm. Spirituality and science both ask the question "Who are we?" but they answer in different ways. Science explores the biological aspect of human life and its evolution, whereas spirituality seeks a meaning and order to existence that is deeper than the one science can provide.

The scientific view has become dominant over the past two centuries, at least in the West. Many are satisfied with its answers and do not feel a need for deeper meaning and order. However this may change when they, for whatever reason, happen to have a spiritual encounter. Authentic spiritual encounter brings the feeling of touching something much larger than oneself and yet paradoxically of being most truly oneself.

The experience of a deeper order does not need to preclude a scientific worldview. Newton and Kepler saw the regularity of the physical world as the handiwork of God and felt they moved closer to the Divine by understanding the natural order. Einstein expressed a related sentiment: "I consider the comprehensibility of the world (to the extent that we are authorized to speak about such a comprehensibility) as a miracle or an eternal mystery ... a priori one should expect a chaotic world which cannot be grasped in any way."[1] He also said, "My religiosity consists in a humble admiration of the infinitely superior spirit that reveals itself in the little that we, with our weak and transitory understanding, can comprehend of reality."[2]

Many see evidence of a deeper meaning and order to come from the opposite direction, from apparent violations of natural law. According to the major Western religious traditions, God occasionally intercedes directly in


*Understanding Intuition*
ISBN 978-0-12-814108-3
https://doi.org/10.1016/B978-0-12-814108-3.00009-9

human affairs and the cosmos. This is believed to occur via transgressions of natural law, such as the parting of the Red Sea, what most (other than Einstein) mean by *miracles*. The idea of divine intervention in the natural order is an important reason why science and religion are often seen as antagonistic.

Evidence for the miraculous—either the existence and (partial) comprehensibility of natural law, or, alternatively, its transgression—may help the conscious mind entertain the idea of an underlying meaning and order to life and the universe. But given that these diametrically opposed arguments both lead to the same conclusion, neither the initial impetus for considering the spiritual stance, nor its ultimate acceptance can hinge on "evidence" alone. Authentic acceptance of an underlying meaning and order, rather than evidence and intellectual argument, rests on another, deeper kind of knowing: on intuition.

## SPIRITUAL INTUITION AND KNOWING WHAT ONE CANNOT KNOW

All different kinds of intuition depend at least in part on knowledge that is *direct* in the sense of being unmediated by the conscious mind, as the proceeding chapters have stressed. Ordinary intuition accounts for much of animal as well as human behavior and knowledge. It begins with bodily and other sensory experience as it interacts with the biological imperative for species survival and the related instinct to seek the well-being of self.

Spiritual intuition also depends on direct knowing. However, it rests on knowledge that lies outside the motivations of the usual self. It seems to reflect the experience of a part of the mind that is tuned differently from the ordinary self, with its conditioned perceptions grounded in survival needs. To put it another way, spirituality may rest on knowing what the ordinary self cannot know. (Of course, this knowledge must be allowed by biology, whether or not it is actively shaped by biological forces.) Spirituality is perhaps most often experienced by the ordinary mind as an intimation of something deeper.

This does not mean that spiritual intuition excludes sensory experience. On the contrary, much of the time the direct knowing of spirituality is intimately associated with sensory experiences. For many, walking in a beautiful natural setting, or perhaps deeply appreciating a flower, may generate a sense of spiritual encounter and the subtle yet enlivened sensibility that accompanies such experiences.

Indeed most experience involves sensation of one sort or another. This implies that something in addition to ordinary sensation marks a given encounter as a spiritual one rather than just a memorable sensual one. Imagination intensified by emotion (sometimes referred to as poetic intensification) is often part of the answer. However I have come to feel that there may be a more direct connection between spirituality and sensation. My proposal will seem radical to some, no doubt. Yet keep in mind that research shows that spiritual experiences, whatever else they may or may not reflect, correspond to real events in the brain.[3,4]

This something else, I submit, reflects an especially subtle and refined quality of sensation. It most often mixes with more ordinary sensation, imbuing it with a numinous, or realer-than-real, quality. Sometimes it can be experienced more or less on its own as an inner, or sometimes as an outer, current. It is an important part of the feeling of personal Presence, which allows us to experience the depth and richness of the moment. Whether the current is experienced in the foreground or background, it seems to carry the knowledge—or the intimation—of deeper order, which is at the core of spiritual experience.

Put another way, authentic spiritual encounter reveals deeper patterns of interconnection. Working via intuition, it allows one to perceive complex patterns of wholeness that can be grasped only dimly if at all at a conscious level.[5] The beauty of the flower or the majesty of the mountains, along with our ability to appreciate them in a realer-than-real way, encodes something about the unity of nature and our place in it.

We may perceive this sense of unity *primarily* as a thought, as an emotional or esthetic feeling, or as an internal or perhaps external current. I suspect that spiritual experience often combines all three. Yet much of the time the more subtle current that underlies the experience goes unrecognized at a conscious level. Of course, we cannot rigorously exclude that spirituality represents, in addition to an altered way of perceiving ordinary reality, an actual external force that can be present to a greater or lesser extent at any one time.[6]

This something-in-addition-to-ordinary-experience that marks spiritual experience in adults may have a prominent role in the way young children perceive their world. After college, I worked as an assistant first-grade teacher for a year. Early in the year, I was astounded by the expressiveness and vitality of the children's paintings. Their special force seemed to depend on the dynamic interconnection of the parts of each composition, along with their vibrant colors. It gave the pictures a vibrating or wave-like feeling. The veteran teacher with whom I worked informed me that this striking vitality would disappear as the children learned to read. Indeed it did as the year progressed. I now feel that the drawings captured at a visual level the special kind of underlying sensing to which I am referring. (This was reinforced by a later intuitive encounter that evoked what I assume was an aspect of my preverbal way of experiencing. While riding in a car one day with my older brother, I experienced him for a short time less as a physical entity and more in terms of the very real-seeming, energetic lines of force connecting us.) The right hemisphere appears to have an important role in the novel interconnected mental apperception pivotal to both intuition and spiritual encounters.[7]

## DIFFERENT KINDS OF SPIRITUAL ANSWERS TO "WHO ARE WE?"

The words *spirituality* and *religion* are often used interchangeably. Yet religion generally implies adherence to specific doctrines. I prefer to use spirituality as a more inclusive term for any activity that evokes an experience of a meaning and order that transcends the material world known by science. I will consider three broad ways of understanding this greater meaning and order. They are at the same time three different answers to the question "Who are we?"

### Dualism

Dualism is the position most often promulgated by religious training in the West. The religious view that a nonmaterial force can mediate deeper meaning leads to the understanding that there are two kinds of substances in the world: spirit and matter. In the 17th century, Descartes codified this dual view of the cosmos in what has come to be known as *Cartesian dualism*. Humans are composed of both *res cogitans*, the thinking thing, and *res extensa*, the extensive and thus material thing.

In his book *Rocks of Ages*, the late paleontologist Steven Jay Gould argued that dualism is the proper stand for contemporary humans.[8] He framed what he called the *magisterium of science* and the *magisterium of religion* as separate yet overlapping areas of human inquiry, each with its own sphere of meaning making and method of investigation. Gould acknowledged that such a separation might not hold in the end. Yet he was provisionally willing to accept the hypothesis that there are two different approaches to answering the question of "Who are we?"

## Mysticism

Mysticism provides another answer; it holds that the ultimate reality is non-physical. Instead it is undifferentiated and without boundaries—the unity of all that is. In some versions of mysticism, it is called *the void*, because it is empty of the illusion of separateness, and in other versions it is called perfect love. Many mystical traditions teach that a number of concentric energetic levels radiate out from this core, generally 7 to 12. The energetic manifestation of the Absolute, which infuses all levels, appears in a less refined state in each successively lower one. The outermost ring, the material world, receives only coarse influences from the Absolute. Various grades of consciousness occupy the levels in between.

According to the mystical perspective, the notion that people are separate and that their consciousness are separate is an illusion. The ego and the conceptual mind keep the psyche tuned to the material level and prevent it from becoming aware of this essential unity. For those on the mystical path, the task is to experience the Absolute at increasingly higher levels.

Mysticism in the West is sometimes associated with unbridled imagination, such as the visions of Blake. Yet for most people, great mental discipline is required to free the mind from its habitual conceptual orientation. Mystical training often emphasizes what has been called *inner science*, or the technology of inner states, which means the centered exploration of mental experience through meditation and other practices.[9] Eventually this can lead to awareness of more refined levels of consciousness, which are at the same time experiences of higher spiritual realms.

Different strands of mysticism diverge somewhat in emphasis and methods, although the basic message and goals are all similar. All answer the question "Who are we?" by saying that we carry the Absolute within. Disciples can experience more and more of what this means through specific practices leading to increasingly refined levels of consciousness and intuition. The assumption of lack of separateness can also, of course, explain intuitive phenomena that have so far been resistant to scientific exploration, such as precognition and extrasensory perception.

## Emergence

In contrast to the mystical claim that lower mental states devolve from higher ones, emergence emphasizes that higher mental states evolve from lower ones. It is at the core of Darwinian evolution, which is based on the idea that higher organisms evolved from lower ones. Emergence offers a potential meeting point for scientific and spiritual perspectives on the question "Who are we?"

Material, or secular, emergence provides a holistic alternative to the traditional reductive approach of science that tries to comprehend the cosmos by looking at increasingly basic components of matter. From the reductive perspective, biology is entirely governed by chemistry, chemistry by physics, and physics by particle physics. Emergence, in contrast, stresses that higher levels of organization can yield properties that cannot be predicted from lower levels.

Consider consciousness. Neurons create consciousness. However studying individual neurons or even groups of neurons and their interactions is not likely to explain the mystery of consciousness. The complexity of the organization of the whole adds something that is irreducible to the parts. To understand consciousness fully, we probably will have to take into account not only the whole brain, but also the body it serves and the world in which the brain and body function.

The secular emergent perspective, like reductionism, considers the basic "stuff" of the universe to be matter. Consciousness along with meaning came into existence only with higher levels of organization of matter. However secular emergence allows for the appearance of higher states of consciousness. Because of the complex connectivity underlying consciousness, it can undergo reorganization under the right conditions to a higher level of functioning, sometimes in response to small changes. Secular emergence, by the same token, at least leaves room for the reality of spiritual experience and its claim of higher meaning and purpose.

Spiritual-based views of emergence, like more secular ones, stress biological evolution. However they tend to see the primary stuff in the universe as having both matter-like and spirit-like qualities. The paleontologist and Jesuit priest Pierre Teilhard de Chardin, who was born in 1881, a year before Darwin died, was one of the first to combine spirituality and evolution in an emergence framework. He held that matter at higher levels of organization reveals its intrinsic consciousness and purpose, which is hidden at lower levels.[10,11]

For Teilhard de Chardin, the germ of the psyche, and thus of meaning and the spirit, lies in an *interiority* that is inherent in matter. Increases in interiority, which correspond to increases in internal complexity, come about through the organizational potential of the kind of energy that he viewed as driving the evolution of organic life. It brings radical new capabilities, such as the ability of a cell membrane to delineate inside and outside, and the ability of the alimentary track to bring the outside in. Eventually it brings consciousness with its ability to delineate awareness of an inside and an outside, and also to bring an interpretation of the outside in (and, as Chapter 7 suggests, also the other way around). In effect, consciousness allows the universe to become aware of itself.

Consciousness also eventually brings the ability to experience this organizing energy as it points toward even higher levels of organization. In the late 1930s, Teilhard de Chardin predicted that sentient life on the surface of the earth would close upon itself and form a unified vibrating band of thought and feeling that blankets and extends out from the planet. Globalization, the invention of the Internet, and discoveries about our genetically determined intersubjectivity (see subsequent discussion) taken together provide evidence in support of his prediction. However time will tell whether another of his predictions, reflecting the strength of his religious belief, was overly optimistic. He felt that the closure he foresaw would ultimately unite the planet in love.

A small but growing number of scientists have views that resonate with some aspects of Teilhard de Chardin's.[12] The influence of Eastern contemplative thought and practice, in conjunction with increasing scientific investigation of meditation[13] and phenomena such as out-of-body[6,14] and near death experience,[15] is beginning to make inroads into the core materialism of contemporary cognitive science. Some view consciousness as external to humans, in addition to a property of the brain. They understand the brain to have evolved the capacity to act as a receiver or transducer of consciousness.

Emergence answers the question "Who are we?" by emphasizing that we are organisms who have emerged from a series of progressive transformations of an elemental substance. Material emergence stresses that the biological portion of our emergence came about through apparently chance mutations interacting with ever-changing environmental conditions. More spiritual emergent perspectives, such as Buddhism, choose a larger frame and stress the interconnection (or the *dependent arising*) of all aspects of nature.[16] Others incorporate mystical elements, which can impart a dualist flavor. And yet others, such as the view of Teilhard de Chardin, emphasize an organizing energy that directs the life process teleologically, or toward a certain end.

All agree that we are organisms who with the experience of consciousness have become aware of our own emergence. With this awareness, emergence continues, but it is necessarily transformed. The gift of human consciousness brings the ability to reflect on the question "Who are we?"

## THE COLLECTIVE JOURNEY

Awareness of self brings the capacity to recognize our separateness and at the same time our interconnection with each other and with the rest of nature. Natural selection, which depends on competition for limited resources, has allowed our species with its superior intelligence to dominate the planet. To a large extent, we have tamed the natural world and now employ it for our own purposes. Yet the same set of skills that served us as a species in the previous phase of evolution has the potential to destroy our civilization in the current one.

In his 1973 book *The Survival of the Wisest,* Jonas Salk, the biologist who created the initial polio vaccine, referred to the current stage of evolution as *metabiological evolution.*[17] In contrast to Teilhard de Chardin, who emphasized its upside, Salk warned of its pitfalls. Using the growth rate of bacteria as a model, he suggested that for the human collective to survive, the principles that support growth at low density when resources are plentiful have to give way to self-regulatory principles when they become limited. For bacteria, he noted, these mechanisms are encoded in the genome. Without positing an answer, he asked whether humans might have similar self-regulatory mechanisms.

Metabiologic evolution, according to Salk, is characterized by a dynamic and necessary tension between *ego,* with its focus on separateness, defense, competition, and acquisition, and *Being,* with its focus on interconnection, cooperation, and self-transcendence. The conflict between the two plays out in each of us personally, as the next section explores. However, it is probably easier to see as it is dramatized on the national and international stages. Consider, for example, the inherent tension between the efforts to keep the International Space Station afloat, on the one hand, and the urge of some to put defensive weapons in space, on the other.

As a species, our inherent competitive urge and need to dominate others has always been tempered by our social needs and the strategic advantage of economic and other kinds of cooperation. Brain recording studies reveal that our interconnection goes even deeper. We are also an intersubjective species. Intersubjectivity is inherent to our genetic makeup and our evolutionary heritage.

We are hardwired to mirror the emotions of others and, moreover, to experience them much as we do our own. The title of one scientific paper on emotional mirroring brings this home particularly poignantly: "Both of Us Disgusted in My Insula: The Common Neural Basis of Seeing and Feeling Disgust."[18] Even seeing a picture of stranger in a painful situation, such as getting a door slammed on his or her hand, or reading about it,[19] stimulates many of the same regions of the brain as if it had happened to us.[20] Moreover, empathy is not unique to humans; it is a highly conserved trait in evolution.[21]

The impact of non–self-derived emotions is much weaker than one's own; nonetheless we are physiologically designed to get under each other's skin. This is true not only of intimate associates but also of complete strangers. Moreover, emotions we do not feel at a conscious level can affect our body, our mood, and behavior.[22] The same is likely true of the emotions of others that we record at an unconscious level. In answer to Salk, it seems that we do have at least one hardwired regulatory mechanism that functions without our awareness to temper some of the more aggressive and exploitative aspects of our nature.[23] Of course, whether this and whatever else interconnects us will be enough to sustain us through the current changes occurring to our planet and civilization remains uncertain.

The research on mirroring explains something a wise rabbi once said that greatly mystified me at the time and grounds it in brain function: "To sin against another is its own punishment." What goes around comes around not only later, but immediately as well. When we harm another, at some level we also register the emotional pain we cause. To barricade this largely hidden emotional discomfort, we experience a little less fully. Without recognizing it, we move further away from awareness of ourselves and our capacity to experience deeply. These barricaded feelings can account for some of the trauma soldiers feel when they return home, as well as the healing power of remorse.

Given our 6 degrees of separation, if we influence everyone we come in contact with below their awareness, and we are also influenced by them, eventually everyone is indirectly influencing everyone else. At a certain density of exposure, we form a complex interconnected system coupled by our unconscious intersubjective experience as it resonates throughout the whole. With the work on emotional mirroring, the idea of humanity as a whole has become biologically grounded, not just a way of thinking. Like any complex interconnected whole, it has a reality that is greater than, as well as separate from, any of its component parts.

Our involuntary interconnection can account for the wonderful feeling of human electricity we sometimes experience in densely populated urban areas such as New York City, and also the relaxation and peace we feel in the wilderness. These background feelings generally influence us below the level of awareness, but we sometimes pick up their echo at a conscious level in what we might call a "vibe." This subtle sense of the collective is part of what we register in the experience of personal Presence, when we are particularly attuned to the depth of the moment.

Of course, we participate in many collectives at the same time. Intersubjectivity is strongest with those with whom we are closest. Their emotions affect us directly, as well as through the reverberating echo of the whole. The idea of a weak connecting force that joins the experience of people not directly connected to each other into emergent groups, and ultimately into humanity as a whole, is analogous to Walter Freeman's model for brain function,[24] discussed in Chapter 4.

To review briefly: Neurons, the principal cells of the brain, make direct electrical connections with each other. They are also indirectly connected through electrical echoes of these connections that flow freely in the intercellular matrix, or media, when one neuron transmits information to another. These currents combine in the intracellular media and make neighboring cells more likely to fire. When summed over large numbers of neurons and groups of neurons, they generate the brain waves revealed by the electroencephalogram.

Much of the time, these extracellular currents are only a weak force. However in the presence of biological reinforcement (via neuromodulator molecules released at the behest of the emotional/motivation system) they become magnified and can entrain the individual cells in a tissue so that they begin to function in a coordinated way. These enhanced extracellular currents, according to Freeman, are also what allow different tissues within a region of the brain, as well as the brain itself, to function as a coordinated whole.

Freeman's work suggests that what we perceive below awareness, and thus what might come to consciousness, reflects a complex hierarchy of patterns of extracellular currents from various regions of the brain, all of which feed back on each other. These patterns are largely governed by past experience as well as by current emotional needs and arousal level. They reflect less the external situation than the unconsciously perceived *meaning* the situation has for the individual at that point in time. For example, when putting things away in the pantry after a dinner ending with a rich dessert, we might not register the box of chocolates that might become more salient a number of hours later. However, after a rough day that leaves us feeling in need of all the comfort we can get, even right after dessert we might note the chocolates' presence for a later treat.

Just as individual perception depends on signals that sum in the intracellular media, I propose that collective perception depends on largely unconscious signals that pass back and forth between us. Especially when amplified by important events, these signals allow groups of people, nations, and, on occasion, humanity, to perceive as a whole. The human collective is a collective of brains (and their bodies), which in turn feeds back to the emotions of individuals—and around it goes. The largely unconscious influences that result from our emotional interconnection join with the other, more personal emotional factors that govern our mood, what we attend to, and the choices we make.

Much of the time the influence of the collective is slight, at best subtly nudging feelings and decision making in a certain direction. However, on occasion it can become strong enough to overwhelm other internal emotional signals and control conscious perception and behavior directly. This accounts for what occurs during riots, certain religious rituals, and periods of war fever—whenever groups of people feel and act in ways uncommon to them.

The potential effect of the collective is greatly amplified by modern communication technology, our collective intercellular medium. In addition to connecting us more and more at an informational level, technology is also enhancing our interconnection at an emotional level. This was evident as many of us watched the millennium celebration ripple with such beauty and grace across the globe, and also when we shared the tragedies of 9/11, the flooding of New Orleans and many other cities, numerous mass shootings, as well as many other heartbreaking events.

With individual brains, it is possible for a small group of cells with the support of the emotional system to dominate neural activity and reorganize the whole brain in a given direction. The same is true with the collective brain. Leadership depends in part on the ability to reinterpret the collective to itself to harness its power. Leaders who are skilled at reading the collective and galvanizing its largely unexpressed underlying emotional content can sometimes with only a small initial following mold a critical mass of the population to their vision.

Statesmen are leaders who shape the collective to serve goals that resonate with Being, such as Franklin D. Roosevelt's "We have nothing to fear but fear itself" or John F. Kennedy's "Ask not what your country can do for you; ask what you can do for your country." In contrast, some leaders shape the collective to resonate with goals primarily associated with the ego and its tendencies. George W. Bush championed the idea of the "ownership society." The current administration carries the ego-associated view, with its focus on separateness, defense, competition, and acquisition, to an unparalleled extent.

Because our receptivity at this level is involuntary, we are all participants in the common mind whether or not we approve. Even when we hold views that oppose those of the common mind, it is important to become conscious of its potential subtle pull on our own psyche. Only by sensing its draw is it possible to speak out *effectively* against its influence on others. Instead of judgment, this requires addressing the emotional concerns and shared human vulnerability behind the common stance in a way that invokes the countervailing strength of Being.

The skills that won us our hegemony as we emerged as a species and later organized ourselves into competing nations no longer support us. We are increasing interconnection at many different levels. We now all breathe the same air and drink from the same well. Our collective assets, be they our shared environment and other resources, our intersubjective experience, or the genes of our progeny, are the concern of all citizens of the globe. We have a responsibility to manage our collective assets for the benefit of the many rather than the few, and for future generations.

If our civilization is to endure, we must move toward a more authentic collective identity grounded in something more spiritually sustaining as well as materially forward-looking than the accumulation of wealth and power and the fostering of ever-accelerating consumer demand. Our own quality of life and the quality of life of those who come after us, and perhaps their survival, depends on this. The future will not take care of itself unless we lend it our concern. As Jonas Salk suggested in the 1980s, we need to ask "What kind of ancestors will we have been?"[25]

Salk posed his question when the genetic revolution was in its infancy. The issues surrounding genetic engineering are more pressing today. We have more than enough knowledge to allow us to tinker with the human genome in a directed way. Yet we do not have and may never have enough knowledge to do so in a fully predictable way. Small changes in the genome may have far-reaching effects, as is true in all complex interconnected systems.

The internal complexity of our genetic blueprint is highlighted by the fact that it has fewer coding genes than the rice plant.[26] At the same time, we do not know the function of most of the 95% of our DNA that does not encode genes. Originally considered "junk DNA," we now know it contains highly conserved regions,[27] encodes some important regulatory functions,[28] and has had a role in evolution.[29] Moreover, the human organism is a complex system embedded in a larger complex system, organic life, which is embedded in an even larger complex system, the earth—and of course it does not stop there.

Yet it is not in our nature to step back from the project of understanding ourselves and the cosmos at ever-deeper levels. We seek knowledge for its own sake as well as to alleviate suffering and increase well-being. Only the deeper force associated with intuition can guide us, given the overwhelming complexity and high stakes, as we try to "thread the needle" and become an active force in forging who we are.

To lend these important issues our concern does not mean having all the answers. To the contrary, one of the most important things we can do is to acknowledge as much of the inherent complexity of the problems we face as we can, and to acknowledge when we are uncertain how to proceed. Opening to intuition, which is the best source of guidance at such times, requires a spirit of deep concern along with the ability to acknowledge not knowing. Admitting the limitation of our knowledge and vision, in addition to encouraging collective discussion, allows us to access the resources of the unconscious mind more fully. As we step back from past attitudes and solutions, we are more able to engage this deeper knowledge that can carry the wisdom of "Who we are" as it emerges from our collective psyche.

## THE PERSONAL JOURNEY

Salk likened Being, which encodes the essence of being human as well as its possibilities, to our genetic system. Correspondingly he likened ego, which interacts with the physical and social environment, to the body, or the somatic system. The analogy makes the point that human functioning is governed to a large extent by unconscious forces—whether or not it was Salk's intent. We have control over neither our genetic programs nor most of structures and physiological mechanisms that govern the body.

A great deal of mental life that was before assumed to occur consciously is now known to occur below awareness, as the previous chapters have stressed. Much of our experience is governed by what I call the intuitive or implicit self, which shapes our perceptions below the level of awareness and hands us the ready-made world we experience. It automatically sets most of our goals, decides what we pay attention to, and marshals memories and other cognitive resources before the conscious mind is involved.

One component of this hidden, or implicit, self is governed by past experience, especially early experience, as it interacts with our genetic proclivities. The conditioned self, or survival self, part of what Salk called the ego, is to a large extent responsible for our dealings with the external world. It can be a source of intuition about survival and worldly well-being. For example, one person might be particularly good at seeing potential risk because below awareness their conditioned self focuses their attention on low-level clues of danger. Another might be sensitive to opportunities others cannot perceive because below awareness their conditioned self focuses their attention on low-level clues of potential success. By the same token, because its perceptions tend to be partial or skewed, the conditioned self is a frequent source of distortion.

The intuitive or implicit self also has another, unconditioned part: Being. This wise inner ally is the source of our experience on occasions when we escape the distorting potential of past experience and feel ourselves to be most fully who we are. It is also the source of more profound intuition. This truth-seeking part of us, as the last chapter mentioned, is also sometimes called the Self.[30]

The conditioned self and the Being are mutually interdependent and should function in a tightly entwined manner, just as the genetic and somatic systems do. However the conditioned self is also shaped by the ego needs of important others as well as by cultural forces, in conjunction with the biologically grounded mechanisms of seeking pleasure and avoiding pain. The goals of the conditioned self along with its ways of construing reality often come to differ from, and sometimes may even oppose, those of Being, or the higher Self.

This means that even if we are skillful at getting what we think we want, something deeper within us may remain unsatisfied. Eventually, some come to recognize that nothing external can fill the deeper needs associated with Being, and they begin to explore the spiritual side of life. Others may turn to spirituality or religion because their conditioned self prevents them from finding the material or worldly success with which to camouflage deeper yearnings.

Being is not necessarily opposed to ordinary pleasure or even to worldly success. However its primary goal is a sense of authenticity and wholeness. It seeks to harmonize the deeper meaning-making needs of the psyche with its legitimate worldly needs. Especially in adulthood, it works to align all aspects of the mind and body that govern attention and behavior. As it becomes stronger, it also becomes an increasingly conscious force.

Being, or the higher Self, often manifests as the experience of *Presence*, which is associated with a deeper and more satisfying kind of well-being. It tends to accompany experiences of profound intuition (even when they are painful), and also awareness of the simpler truth of the current moment. The feeling of well-being it brings is akin to the equanimity of Buddhist *nonattachment*.

Presence is the paradoxical experience of being deeply seated in oneself even as one feels acutely interconnected with (or perhaps immersed in) something much larger than oneself. It is, for example, what directs a rabbi or pastor to stay focused at the end of a sermon and hold the congregation's attention for a moment or two until the deeper energy of the collective shifts to what is coming next. By moving awareness from the surface of the current moment to its deeper currents, it may alter habitual assumptions about what is *figure* and what is *ground*.

The enhanced consciousness of Presence brings with it the ability to be open to the fullness and depth of the moment. It is a deeply personal yet also impersonal feeling of spaciousness. Most of the time, the gaps between our words and between our sentences are filled with internal mental chatter. Presence allows us to savor the richness of the empty spaces between our words and our sentences and between a question and its answer. It alters the experience of the moment from something ever-fleeting, what I will call its *extensive dimension*, to that of a deep energetic container, its *intensive dimension*. It senses and savors meaning, sometimes with words, although perhaps more often with the full current of Being.

Presence allows us to stand outside our conditioned associations and emotional reactions and be permeable to the myriad layers of experience that shape the moment. In this sense, it is also akin to the Buddhist concept of *emptiness*, which allows the experience of dependent arising, or the interconnection of all things. By the same token, the mental quiet that accompanies Presence allows tacit knowledge to emerge from the unconscious relatively unadulterated by the usual considerations and defenses that distort our perception. By focusing on the depth of the moment, Presence allows us to be more open to what is really of value and to our own deeper knowledge.

The profound spaciousness and ease associated with Being is generally won only with a certain amount of discomfort or pain, just as the pearl in the oyster is a response to the irritation of sand. The difficulties life inevitably strews on all our paths illuminate the limitations of our specific conditioned ways of perceiving. The failure of our habitual ways of construing reality can help us align more deeply with Being's goal of wholeness and attunement to the deeper currents of the moment that run through us and around us.

The more distressing and significant the event or events are that occur in our lives, the more difficult it is to hold on to our habitual meaning-making framework. In turn, the greater the need is to find a larger framework of meaning. The crucial change in perspective occurs, I believe, when we can even hesitantly frame a significant and deeply painful experience not as senseless suffering, but instead as a *spiritual challenge*. By bracketing the experience as a challenge in meaning making, we begin implicitly, if not explicitly, to examine the question at the root of all higher meaning making: "Who Am I?" Held in such a context, life's difficulties become opportunities that lead more deeply into the mystery of Being: into the intensive dimension of time and existence.

One of the things that can happen during this questioning process is that the conscious mind begins to realize that the version of reality that the survival self hands to it is not the only possible one. In standing back from current perceptions and looking at them and the discomfort they bring, one can take the initial steps toward recognizing the systematic way the conditioned self slants its version of reality. I had a lesson in the potential power of using difficult challenges to foster spiritual growth in the process of writing this book.

Another book on unconscious cognition and intuition was published and became a runaway best seller, just when I thought I was close to finishing this project—which had already taken some years.[31] It used the idiom "a blink of the eye" as its central metaphor but did not cite the 1997 article in which I had previously published this striking characterization of intuition, along with the unusual experience in which it came to me.[32] All of my subsequent efforts at the time to ensure publication of my book were frustrated.

I did recognize, at least intellectually, that the roadblock was likely temporary and my book eventually would find its audience. I also realized that a popular book on unconscious cognition and intuition by a well-known author was actually a positive step. Furthermore, I saw that the popularization of the blink of the eye metaphor for intuition would in the end help readers of my book get beyond its topmost meaning to the many more profound messages it encodes. Nonetheless I continued to feel terrible.

While sweeping up the kitchen one night after dinner, I turned and faced the apparent paradox of continuing to feel unvalued, unseen, and even annihilated, despite my sense that eventually my book would be published. In response, the feeling of despair gave way to an unprecedented feeling of spaciousness and wholeness! Accompanying this was the striking realization that publishing the book was nowhere near as important as gaining more ready access to this wonderful feeling of spaciousness, coherence, and wholeness. I saw that writing the book was a much deeper adventure than producing a published product, which would not likely satisfy the deeper meaning-making needs in me. Perhaps I implicitly realized what I now see at a conscious level: I would be able to publish the book only when it was no longer important to do so in order to validate myself.

As this incident reveals, when we can, even if only tentatively, begin to turn toward and bracket our suffering, one of the things that can happen is that we get a glimpse of our habitual perceptual or meaning-making process itself. Along with this may come a hint of some of the basic human permeability and vulnerability this way of framing reality was in part crafted to protect. Instead of the low self-esteem that usually accompanies self-judgment, recognition of this vulnerability can bring a sense of purity, wholeness, and spaciousness. To the extent that we grasp that we share this underlying permeability and vulnerability with all other humans, we can also recognize at an ever-deeper level that those who may cause us pain are also victims of their own conditioned or survival selves and its distorted way of framing reality. From this seed, compassion can grow.

Of course, using whatever challenges come our way as aids to spiritual growth does not mean we encourage or honor those who either consciously or unconsciously, or even inadvertently, cause us pain. Nor does it necessarily mean we need eschew actions to protect ourselves from exploitation. However taking action too soon to counteract the discomfort can sometimes prevent deeper meaning-making work. If instead we can greet the difficulties we face with even a small part of ourselves as spiritual challenges, as challenges in Being, we may begin to see new ways of perceiving reality and taking care of ourselves. As we move deeper into this transformative process, we might find that the kind of action we would take, or perhaps the feeling that drives it, will change.

I had yet another surprise when I was finishing a close-to-last draft of this chapter.[33] Serendipitously I learned about the work of the Danish philosopher Kierkegaard, who, like Salk, was concerned with the interrelationship of different aspects of the psyche. He wrote extensively about the challenge of opening to authentic Being. "… [T]he progress of becoming must be an infinite moving away from itself in the infinitizing of the self, and an infinite coming back to itself in the finitizing process"[34]

Kierkegaard wove the notion of "the Moment," which resonates strongly with what I have called Presence, throughout his work. A deeply religious man, Kierkegaard saw "the Moment" as, among other things, a brief entry into eternity, with its complete knowledge and vision. As it turns out, the literal meaning of the (old) Danish word for moment, *Øiblikket*, is "a blink of the eye!"

In these moments of spiritual as well as intuitive grace, internal and external influences converge in such a way that a reorganization occurs in the brain/mind. These "blinks of the eye/I" bring one home to Being—to whom one is. Although only temporary, they provide a glimpse of what lies beyond the habits that usually cloud perception and experience.

More and more, we can encourage these moments by aligning ourselves *consciously* with the desire to move beyond the limitations of the survival self and its conditioned view of reality. One prerequisite for this is openness to new, more profound answers to the question "Who Am I?" In contrast to the conditioned self, which is primarily attached its own perpetuation, Being is impartial, unattached, and open to the truth of the moment in all its depth.

To the extent that we are Present, which means deeply permeable to as well as conscious of the current of life in and around us, each moment offers the possibility of unfolding into authenticity and wholeness, of coming home. Like the pastor or rabbi who lets something deeper, more integrated, and emergent than the ordinary mind choose the right moment to continue the service after the sermon, we can let intuition as it emerges from below awareness guide us into the future. To know more deeply, we have to open to not-knowing; likewise, to be truly our self, we have to surrender what most of the time we take to be the self.

The force of intuition is our compass and the compass of our species. It always works through us below awareness, gathering information and guiding our mental activity. However it is largely filtered by the survival self with its partial vision. To the extent that intuition is instead guided by the Being or the Self, it can become a source of wisdom and also enhanced power. It illuminates the path in front of us and helps us learn both as individuals and as a species more and more who we are.

The journey into the mystery of human existence and its unfolding is both a scientific and a spiritual adventure. The two are not necessarily as far apart as they might seem. The impartial openness and commitment to truth of Being shares something with the objectivity of science. Science seeks objectivity through empirical study, whereas Being seeks objectivity by going beyond the habitual perception of the survival self to a deeper substratum of intuition. Yet intuition has a significant role in guiding scientific endeavor. Moreover to the extent that scientific inquiry is guided by the deeper intuition of Being, it also becomes a form of spiritual inquiry. Einstein wrote:

> *A finely tempered nature longs to escape from personal life into the world of objective perception and thought: this desire may be compared with the townsman's irresistible longing to escape from his noisy cramped surroundings into the silence of high mountains, where the eye ranges freely through the still, pure air and fondly traces out the restful contours apparently built for eternity.*[35]

## REFERENCES


1. Einstein A. *Letters to solovine: 1906–1955*. New York: Philosophical Library; 2015. p. 117.
2. Einstein A. In: Dukas H, Hoffman B, editors. *Albert Einstein the human side*Princeton (NJ): Princeton University Press; 1981. p. 66.
3. Newberg A, D'Aquilli EG. *Why god won't go away*. New York: Random House; 2001.
4. Ramachandran VS, Blakeslee S. *Phantoms in the brain: probing the mysteries of the human mind*. New York: Wm. Morrow; 1998. p. 174–98.
5. For an interesting perspective that intersects at some points with the one I give, Galin D. *Rehabilitating the concept of 'spirit' for the non-religious and scientifically minded*. Available from: http://rogerr.com/galin/papers/spirit_1.htm.
6. For an article making a similar point–being able to naturalize a concept–does not mean the naturalized explanation is necessarily complete or correct Metzinger T. Out-of-body experiences as the origin of the concept of the soul. *Mind Matter* 2005;**3**:51–87.
7. Taylor JB. *My stroke of insight*. New York: Vintage; 2008. Taylor, a neuroscientist, suffered a stroke that disabled most left hemisphere function. Her description of her transformed experience of herself and the world resonates with what I saw in the pictures of the children and also what disappeared when they learned to read and write.
8. Gould SJ. *Rocks of ages: science and religion in the fullness of life*. New York: Bantam; 1999.
9. Goleman D, Thurman R, editors. *MindScience: an east west dialogue*. Boston: Wisdom Publications; 1991.
10. The best known of his book is P. Teilhard de Chardin. *The phenomenon of man*. New York: Harper & Row; 1975.
11. For a different spiritual emergent view, by a contemporary philosopher and theologian Clayton P. The emergence of spirit. *Cent Theol Nat Sci Bull* 2000;**20**:3–20.
12. Forman RCK. A watershed event: neuroscience consciousness and spirituality. *J Conscious Stud* 2008;**15**:110–5.

13. Tang Y-Y, Holzel BK, Posner MI. The neuroscience of mindfulness meditation. *Nat Rev Neurosci* 2015;**16**:213–25.
14. Lenggenhager B, Tadi T, Blanke O. Video Egro sum: manipulating bodily self-consciousness. *Science* 2007;**317**:1069–99.
15. Van Lommel P. Near-death experiences: the experience of the self as real and not as an illusion. *Ann NY Acad Sci* 2011;**1234**:19–28.
16. Varela FJ, Thompson E, Rosch E. *The embodied mind: cognitive science and the human experience*. Cambridge (MA): MIT Press; 1991.
17. Salk J. *The survival of the wisest*. New York: Harper & Row; 1973.
18. Wicker B, Keysers K, Plailly J, et al. Both of us disgusted in my insula: the common neural basis of seeing and feeling disgust. *Neuron* 2003;**40**:655–64.
19. Jackson PL, Meltzoff A, Decety J. How do we perceive pain in others: a window into the neural processes involved in emotion. *Neuroimage* 2005;**24**:771–9.
20. Ochsner N, Zaki J, Hanelin J, et al. Your pain or mine? Common and distinct neural systems supporting the perception of pain in self and other. *Scan* 2008;**3**:143–60.
21. Decety J, Bartal IB-A, Uzefovsky F, Knafo-Noam A. Empathy as a driver of prosocial behavior: highly conserved neurobehavioural mechanisms across species. *Philos Trans R Soc Lond B* 2016;**371**(1686):20150077. https://doi.org/10.1098/rstb.2015.0077.
22. Damasio A. *Descartes' error: emotion, reason, and the human brain*. New York: Putnam's Sons; 1994.
23. For an article suggesting that in addition to empathetic simulation mechanisms, some emotional contagion between humans seems to be transmitted directly via chemosensory molecules see Prehn-Kristensen A, Wiesner C, Bergmann TO, Wolff S, Jansen O, et al. Induction of empathy by the smell of anxiety. *PLoS One* 2009;**4**. https://doi.org/10.371/journal.pone.0005987.
24. Freeman W. *How brains make up their minds*. London: Weidenfeld & Nicolson; 1999.
25. Salk J. *Anatomy of reality: merging of intuition and reason*. New York: Columbia University Press; 1984.
26. Sterck L, Rombauts S, Vandepoele K, Rouze P, Van de Peer Y. How many genes are there in plants (... and why are they there)? *Curr Opin Plant Biol* 2007;**10**:199–203.
27. Siepel A, Bejerano G, Pederson SJ, et al. Evolutionary conserved elements in vertebrate, insect, worm, and yeast genomes. *Genome Res* 2005;**15**:1034–50.
28. Hammock EA, Young LJ. Functional microsalellite polymorphism associated with divergent social structure in vole species. *Mol Biol Evol* 2004;**21**:1057–63.
29. Prabhakar S, Visel A, Akiyama JA, Shoukry M, Louis KD, Holt A, et al. Human-specific gain of function in a developmental enhancer. *Science* 2008;**321**:1346–50.
30. This higher part of human nature that seeks authenticity and wholeness has been called many different names, such as *Essence, True Nature*, and the *unlimited Self*. Buddhists feel that there is no ultimate center of self; instead all organization comes from *dependent arising*. Although the view I have presented here is more *essentialist*, it is compatible in many respects with the idea of dependent arising. The conscious aspect of Self, Presence, which is a form of what the Buddhists call *mindfulness*, allows a glimpse of dependent arising.
31. Gladwell M. *Blink: the power of thinking without thinking*. New York: Little Brown; 2005.
32. Isenman L. Towards an understanding of intuition and its importance in scientific endeavor. *Perspect Biol Med* 1997;**40**:395–403.
33. May R. *Existential psychology*. New York: McGraw Hill; 1969. p. XX.
34. Kierkegaard S. The sickness unto death: a christian psychological exposition for upbuilding and awakening. In: Hong HV, Hong EH, editors. *Kierkegaard's writing*, vol. 19. Princeton (NJ): Princeton University Press; 1983.
35. From an address by Albert Einstein entitled "Principles of Research," delivered in 1918 at the Physical Society Berlin in honor of Max Planck's sixtieth birthday. Available from: http://speakola.com/ideas/albert-einstein-principles-of-research-1918.

# INDEX

'*Note:* Page numbers followed by "f" indicate figures, "t" indicate tables.'

## C



## D



## E

## S

## T



## U



## V



## W